Chemical Tests

TEACHER'S GUIDE

SCIENCE AND TECHNOLOGY FOR CHILDREN

NATIONAL SCIENCE RESOURCES CENTER
Smithsonian Institution • National Academy of Sciences
Arts and Industries Building, Room 1201
Washington, DC 20560

NSRC

The National Science Resources Center is operated by the Smithsonian Institution and the National Academy of Sciences to improve the teaching of science in the nation's schools. The NSRC collects and disseminates information about exemplary teaching resources, develops and disseminates curriculum materials, and sponsors outreach activities, specifically in the areas of leadership development and technical assistance, to help school districts develop and sustain hands-on science programs. The NSRC is located in the Arts and Industries Building of the Smithsonian Institution and in the Capital Gallery Building in Washington, D.C.

ISBN 0-89278-704-X

Published by Carolina Biological Supply Company, 2700 York Road, Burlington, NC 27215.
Call toll free 800-334-5551.

Foreword

Since 1988, the National Science Resources Center (NSRC) has been developing Science and Technology for Children (STC), an innovative hands-on science program for children in grades one through six. The 24 units of the STC program, four for each grade level, are designed to provide all students with stimulating experiences in the life, earth, and physical sciences and technology while simultaneously developing their critical thinking and problem-solving skills.

Sequence of STC Units

Grade	Life, Earth, and Physical Sciences			
1	Organisms	Weather	Solids and Liquids	Comparing and Measuring
2	The Life Cycle of Butterflies	Soils	Changes	Balancing and Weighing
3	Plant Growth and Development	Rocks and Minerals	Chemical Tests	Sound
4	Animal Studies	Land and Water	Motion and Design	Electric Circuits
5	Microworlds	Ecosystems	Food Chemistry	Floating and Sinking
6	Experiments with Plants	Measuring Time	The Technology of Paper	Magnets and Motors

The STC units provide children with the opportunity to learn age-appropriate concepts and skills and to acquire scientific attitudes and habits of mind. In the primary grades, children begin their study of science by observing, measuring, and identifying properties. Then they move on through a progression of experiences that culminate in grade six with the design of controlled experiments.

Sequence of Development of Scientific Reasoning Skills

Scientific Reasoning Skills	Grades					
	1	2	3	4	5	6
Observing, Measuring, and Identifying Properties	♦	♦	♦	♦	♦	♦
Seeking Evidence Recognizing Patterns and Cycles		♦	♦	♦	♦	♦
Identifying Cause and Effect Extending the Senses				♦	♦	♦
Designing and Conducting Controlled Experiments						♦

The "Focus–Explore–Reflect–Apply" learning cycle incorporated into the STC units is based on research findings about children's learning. These findings indicate that knowledge is actively constructed by each learner and that children learn science best in a hands-on experimental environment where they can make their own discoveries. The steps of the learning cycle are as follows:

- Focus: Explore and clarify the ideas that children already have about the topic.

- Explore: Enable children to engage in hands-on explorations of the objects, organisms, and science phenomena to be investigated.

- Reflect: Encourage children to discuss their observations and to reconcile their ideas.

- Apply: Help children discuss and apply their new ideas in new situations.

The learning cycle in STC units gives students opportunities to develop increased understanding of important scientific concepts and to develop better attitudes toward science.

The STC units provide teachers with a variety of strategies with which to assess student learning. The STC units also offer teachers opportunities to link the teaching of science with the development of skills in mathematics, language arts, and social studies. In addition, the STC units encourage the use of cooperative learning to help students develop the valuable skill of working together.

In the extensive research and development process used with all STC units, scientists and educators, including experienced elementary school teachers, act as consultants to teacher-developers, who research, trial teach, and write the units. The process begins with the developer researching the unit's content and pedagogy. Then, before writing the unit, the developer trial teaches lessons in public school classrooms in the metropolitan Washington, D.C., area. Once a unit is written, the NSRC evaluates its effectiveness with children by nationally field-testing it in ethnically diverse urban, rural, and suburban public schools. At the field-testing stage, the assessment sections in each unit are also evaluated by the Program Evaluation and Research Group of Lesley College, located in Cambridge, Mass. The final editions of the units reflect the incorporation of teacher and student field-test feedback and of comments on accuracy and soundness from the leading scientists and science educators who serve on the STC Advisory Panel.

Major support for the STC project has been provided by the National Science Foundation, the John D. and Catherine T. MacArthur Foundation, the U.S. Department of Defense, the Dow Chemical Company Foundation, and the U.S. Department of Education. Other contributors include E. I. du Pont de Nemours & Company, the Amoco Foundation, Inc., and the Hewlett-Packard Company.

Acknowledgments

Chemical Tests was developed and written by Wendy R. Binder in collaboration with the STC development and production team. (Several activities were adapted from the unit *Mystery Powders*, which was developed by the Elementary Science Study Project in 1969.) The unit was edited by Lynn A. Miller and Linda Harteker and illustrated by Max-Karl Winkler and Catherine Corder. It was trial taught in Watkins Elementary School of the Capitol Hill Cluster Schools in Washington, DC.

The technical review of *Chemical Tests* was conducted by:

Andrew R. Barron, Professor, Department of Chemistry, Harvard University, Cambridge, MA

Alan Mehler, Professor, Department of Biochemistry, Howard University Medical School, Washington, DC

The unit was nationally field-tested in the following school sites with the cooperation of the individuals listed:

Cleveland School District, Cleveland, OH
Coordinator: Bill Badders, Science Coordinator
Lourdes Gonzalez, Teacher, Joseph Landis Elementary School
Pam Poveroni, Teacher, Charles Dickens Elementary School
Ted Shaft, Teacher, Miles Park School

Santa Rosa School District, Santa Rosa, NM
Ken Livingston, Science Coordinator
Alice Booky, Teacher, Santa Rosa Elementary School
Lorraine Madrid, Teacher, Santa Rosa Elementary School
Flora Ortiz, Teacher, Santa Rosa Elementary School

Ojibwa Indian School, Belcourt, ND
Kathy Henry, Science Department Head
Mary-Jo Hannesson, Teacher
Charmane Johnston, Teacher

Kingfield School District, Kingfield, ME
Jeanne Tucker, Science Coordinator
Deborah Knapp, Teacher, Phillips Middle School
Sue Marden, Teacher, Kingfield Elementary School
Pamela Mauzaka, Teacher, Strong Elementary School

Clifton Elementary School, Clifton, VA
Jody Hepner, Teacher

The NSRC also would like to thank the following individuals for their contributions to the unit:

Ann Benbow, Editor, Education Division, American Chemical Society, Washington, DC
L. J. Benton, Instructional Materials Specialist, Fairfax, VA
Debby Deal, Curriculum Consultant, Clifton, VA
Rayna D. Green, Director, American Indian Program, Department of the History of Science and Technology, National Museum of American History, Smithsonian Institution, Washington, DC
Joe Griffith, Director, Hands-on Science Center, Science in American Life, National Museum of American History, Smithsonian Institution, Washington, DC
Gloria F. Henderson, Vice Principal, Watkins Elementary School, Washington DC
Theresa Hill, Science Teacher, Watkins Elementary School, Washington, DC
Eric Long, Staff Photographer, Office of Printing and Photographic Services, Smithsonian Institution, Washington, DC
Jan Loveless, Educational Consultant, Midland, MI
Barbara Lucas, Teacher, Watkins Elementary School, Washington, DC
Mary Ellen McCaffrey, Photographic Production Control, Smithsonian Institution, Washington, DC
Dane Penland, Chief, Special Assignments/Photography Branch, Office of Printing and Photographic Services, Smithsonian Institution, Washington, DC
Laura Pierce, STC Assistant (1990-92), National Science Resources Center, Washington, DC
Elly Uehling, Teacher, Wakefield Forest Elementary School, Fairfax, VA
The librarians and staff of the Central Reference Service, Smithsonian Institution Libraries, Washington, DC

STC Advisory Panel

Peter P. Afflerbach, Associate Professor, National Reading Research Center, University of Maryland, College Park, MD

David Babcock, Director, Board of Cooperative Educational Services, Second Supervisory District, Monroe-Orleans Counties, Spencerport, NY

Judi Backman, Math/Science Coordinator, Highline Public Schools, Seattle, WA

Albert V. Baez, President, Vivamos Mejor/USA, Greenbrae, CA

Andrew R. Barron, Associate Professor of Chemistry, Harvard University, Cambridge, MA

DeAnna Banks Beane, Project Director, YouthALIVE, Association of Science-Technology Centers, Washington, DC

Audrey Champagne, Professor of Chemistry and Education, and Chair, Educational Theory and Practice, School of Education, State University of New York at Albany, Albany, NY

Sally Crissman, Faculty Member, Lower School, Shady Hill School, Cambridge, MA

Gregory Crosby, National Program Leader, U.S. Department of Agriculture Extension Service/4-H, Washington, DC

JoAnn E. DeMaria, Teacher, Hutchison Elementary School, Herndon, VA

Hubert M. Dyasi, Director, The Workshop Center, City College School of Education (The City University of New York), New York, NY

Timothy H. Goldsmith, Professor of Biology, Yale University, New Haven, CT

Charles N. Hardy, Assistant Superintendent for Instructional Services, Highline Public Schools, Seattle, WA

Patricia Jacobberger Jellison, Geologist, National Air and Space Museum, Smithsonian Institution, Washington, DC

Patricia Lauber, Author, Weston, CT

John Layman, Professor of Education and Physics, University of Maryland, College Park, MD

Sally Love, Museum Specialist, National Museum of Natural History, Smithsonian Institution, Washington, DC

Phyllis R. Marcuccio, Assistant Executive Director of Publications, National Science Teachers Association, Arlington, VA

Lynn Margulis, Professor of Biology, Department of Botany, University of Massachusetts, Amherst, MA

Margo A. Mastropieri, Co-Director, Mainstreaming Handicapped Students in Science Project, Purdue University, West Lafayette, IN

Richard McQueen, Teacher/Learning Manager, Alpha High School, Gresham, OR

Alan Mehler, Professor, Department of Biochemistry and Molecular Science, College of Medicine, Howard University, Washington, DC

Philip Morrison, Professor of Physics, Emeritus, Massachusetts Institute of Technology, Cambridge, MA

Phylis Morrison, Educational Consultant, Cambridge, MA

Fran Nankin, Editor, SuperScience Red, Scholastic, New York, NY

Harold Pratt, Senior Program Officer, Development of National Science Education Standards Project, National Academy of Sciences, Washington, DC

Wayne E. Ransom, Program Director, Informal Science Education Program, National Science Foundation, Washington, DC

David Reuther, Editor-in-Chief and Senior Vice President, William Morrow Books, New York, NY

Robert Ridky, Associate Professor of Geology, University of Maryland, College Park, MD

F. James Rutherford, Chief Education Officer and Director, Project 2061, American Association for the Advancement of Science, Washington, DC

David Savage, Assistant Principal, Rolling Terrace Elementary School, Montgomery County Public Schools, Rockville, MD

Thomas E. Scruggs, Co-Director, Mainstreaming Handicapped Students in Science Project, Purdue University, West Lafayette, IN

Larry Small, Science/Health Coordinator, Schaumburg School District 54, Schaumburg, IL

Michelle Smith, Publications Director, Office of Elementary and Secondary Education, Smithsonian Institution, Washington, DC

Susan Sprague, Director of Science and Social Studies, Mesa Public Schools, Mesa, AZ

Arthur Sussman, Director, Far West Regional Consortium for Science and Mathematics, Far West Laboratory, San Francisco, CA

Emma Walton, Program Director, Presidential Awards, National Science Foundation, Washington, DC, and Past President, National Science Supervisors Association

Paul H. Williams, Director, Center for Biology Education, and Professor, Department of Plant Pathology, University of Wisconsin, Madison, WI

STC Development and Production Team

Joyce Lowry Weiskopf, Project Director
Wendy Binder, Research Associate
Edward V. Lee, Research Associate
Katherine Stiles, Research Associate
Katherine Darke, Program Assistant
Carol O'Donnell, Program Consultant

Kathleen Johnston, Publications Director
Dean Trackman, Managing Editor
Lynn A. Miller, Writer/Editor
Max-Karl Winkler, Illustrator
Heidi Kupke, Publications Technology Specialist
Laura Akgulian, Writer/Editor Consultant
Linda Harteker, Writer/Editor Consultant
Dorothy Sawicki, Writer/Editor Consultant
Lois Sloan, Illustrator Consultant
Martha Vaughan, Illustrator Consultant

NSRC Administration

Douglas Lapp, Executive Director
Sally Goetz Shuler, Deputy Director
Karen Fusto, Administrative Officer
Diane Mann, Associate Administrative Officer
Kathleen Holmay, Public Information Consultant
Gail Greenberg, Executive Administrative
   Assistant
Karla Saunders, Administrative Assistant

STC Evaluation Consultants

George Hein, Director, Program Evaluation and
   Research Group, Lesley College
Sabra Price, Senior Research Associate, Program
   Evaluation and Research Group, Lesley College

Contents

Goals for *Chemical Tests*

In this unit, students investigate the properties of a variety of common household chemicals. From their experiences they are introduced to the following science concepts, skills, and attitudes.

Concepts

- Common household chemicals have different physical and chemical properties.

- Chemicals undergo changes in form, color, or texture when they are mixed together, separated, or heated.

- Some chemicals can be identified by their interaction with water, vinegar, iodine, red cabbage juice, and heat.

- Different types of mixtures, such as solutions or suspensions, are created when solids are combined with water.

- Evaporation and filtration are methods for separating mixtures of solids and liquids.

- Some chemicals can be classified as acids, bases, or neutral substances by their reactions with red cabbage juice.

Skills

- Observing and describing properties of materials.

- Learning to perform different physical and chemical tests.

- Predicting, observing, describing, and recording results of tests.

- Analyzing and drawing conclusions from the results of tests.

- Comparing and contrasting test results to define the properties of household chemicals so they can be identified.

- Supporting conclusions with reasons based on experiences.

- Communicating results and reflecting on experiences through writing and discussion.

- Applying previously learned knowledge and skills to new situations to solve a problem.

- Reading to enhance understanding of chemistry concepts.

- Developing proper lab techniques to ensure safety and avoid contamination.

Attitudes

- Developing interest in and enthusiasm toward exploring and investigating properties of chemicals.

- Recognizing the importance of guidelines for experimentation.

- Developing an awareness of the importance of chemicals in our lives.

- Developing an appreciation for the safe handling of chemicals.

Unit Overview and Materials List

Chemistry. It's not confined to a laboratory filled with test tubes—or a lecture at a university. Chemistry is the study of the world around us, of chemicals and how they interact. Everything is made of chemicals, from the pencils we write with to the oceans where we swim. We are made of chemicals. When we eat or breathe, chemical reactions take place inside us.

Many students have been amazed to realize that chemicals are, in fact, all around us. And, they are surprised that chemistry is a topic they can explore in their own classroom. To study chemistry is to make sense of the world. In a way, it's like solving a mystery.

In *Chemical Tests,* a 16-lesson unit designed for third-graders, students begin to use critical thinking skills to solve mysteries such as "What are the identities of five unknown solids?" By conducting a variety of physical and chemical tests, students explore some concepts basic to general chemistry: physical and chemical properties and how to describe them, and changes that may occur when different solids and liquids are mixed together or separated. As a result, students are introduced to solubility, filtration, evaporation, crystallization, and acids, bases, and neutrals. Along the way, they also develop essential skills: observing, recording, questioning, analyzing, and drawing conclusions.

At this age, many children increasingly can infer and predict on the basis of experiences, and they are beginning to support their conclusions both verbally and in written form. Using knowledge gained through observation and experimentation, third-graders also are starting to be able to solve the problems posed in this unit.

The unit begins with a pre-unit assessment lesson in which students share what they think about chemicals and what they would like to learn and gain from their first experience observing and describing an unknown material. In Lesson 2, students encounter the mystery of their five unknown solids (sugar, alum, talc, baking soda, and cornstarch), and assemble the tools they will use to gather the information necessary to solve it. They also practice observing and describing skills with known materials before moving on to unknowns in the next lesson.

In the next eight lessons, students perform a series of tests to help them determine some physical and chemical properties of their five unknown solids. In Lesson 3, students begin to investigate and record the physical properties of the five unknowns, first using their unaided senses, and then adding tools to extend and enhance their senses.

By adding water to the unknowns in Lessons 4 to 6, students learn that mixing one substance with another can reveal new properties. They also use filtration and evaporation to separate the combined substances and find that evaporation may lead to the formation of crystals.

By adding vinegar, iodine, and red cabbage juice to the unknowns in Lessons 7 to 9, the class continues to investigate the properties of the solids. By performing these different tests, students discover that mixing one substance with another can produce physical and chemical changes in form, color, odor, or texture. In their final chemical test on the unknown solids in Lesson 10, students observe that adding heat also can produce notable changes.

In Lesson 11, students review, analyze, and discuss the information they have collected to help identify the five unknown solids. In the next lesson, students compare this information to a Chemical Information Sheet, a reliable source of outside data, to discover the unknowns' identities at last. A reading selection illustrates the important role these and other chemicals play in students' everyday lives. Then, in Lesson 13, students articulate the process they have used to solve their mystery and apply their testing, observing, recording, and analyzing skills to a new situation.

Lesson 14 challenges students to go a step beyond identifying one unknown solid as they perform a chemical analysis on a mixture of two unknown solids. Next, students use the red cabbage juice test and a reading selection to learn about three important chemical groups: acids, bases, and neutrals. Students then apply their new knowledge to the five solids they investigated earlier.

Finally, Lesson 16 is an embedded assessment of the skills students have been developing throughout the unit. Here, students face a challenging reversal: instead of using known test liquids (water, vinegar, iodine, and red cabbage juice) to identify an unknown solid, they use the properties of their five original solids (sugar, alum, baking soda, talc, and cornstarch) to identify one of the test liquids.

As students become more aware of the chemicals in their world, they are likely to have a number of questions that cannot easily be answered. It is a good idea to remind them that our knowledge of chemistry is always growing and changing. What you can do is help students learn how to continue to find out for themselves.

Materials List

Below is a list of materials needed for the *Chemical Tests* unit. Please note that the metric and English equivalent measurements in this unit are approximate.

1	*Chemical Tests* Teacher's Guide
15	*Chemical Tests* Student Activity Books
15	plastic cups, 207 ml (7 oz)
71	resealable heavy-duty plastic bags, 1 liter (1 qt)
1	measuring cup, 60 cc (¼ c)
2	boxes of cornstarch, 453 g (1 lb)
1	plastic graduated cylinder, 50 ml (¾ oz)
20	paper trays, 18 x 24 cm (7¼ x 9¼")
15	plastic pails, 4 liter (1 gal)
46	red sticker dots, .5 cm (¼")
46	yellow sticker dots, .5 cm (¼")
46	green sticker dots, .5 cm (¼")
46	blue sticker dots, .5 cm (¼")
46	orange sticker dots, .5 cm (¼")
31	red sticker dots, 2 cm (¾")
31	yellow sticker dots, 2 cm (¾")
31	green sticker dots, 2 cm (¾")
31	blue sticker dots, 2 cm (¾")
31	orange sticker dots, 2 cm (¾")
80	screw top jars, 14 g (½ oz)
80	plastic taster spoons
15	brown paper lunch bags
2	boxes of flat toothpicks
1	roll of wax paper, 23 sq m (75 sq ft)
30	hand lenses
1	pack of assorted colors of construction paper, 23 x 31 cm (9 x 12")
114	plastic dropper bottles, 7 ml (¼ oz)
15	printed "water" labels
15	printed "iodine" labels
15	printed "vinegar" labels
15	printed "red cabbage juice" labels
15	plastic droppers
80	plastic graduated cups, 37 ml (1¼ oz)
80	No. 1 coffee filters
40	petri (evaporation) dishes, 60 x 15 mm (2¼ x 1¾")
2	packs of eight Expo® dry-erase markers
40	aluminum bake cups, 6.35 cm (2½")
1	roll aluminum foil, 7.6 sq m (25 sq ft)
30	wooden clothespins
1	aluminum pie tin, 22.9 cm (9")
2	votive candles
1	box of safety matches
30	sheets of white construction paper, 30 x 46 cm (12 x 18")

63	blank labels
45	plastic soufflé cups, 37 ml (1¼ oz)
45	plastic soufflé cup lids
6	printed "lemon juice" labels
6	printed "ammonia" labels
6	printed "alcohol" labels
6	printed "detergent" labels
6	printed "water" labels
6	printed "vinegar" labels
3	markers: pink, lime green, blue-purple
2	plastic funnels
1	package of food coloring with four colors: red, yellow, blue, green
1	red color-coded stock container of sugar, 1 liter (1 qt)
1	yellow color-coded stock container of alum (ammonium alum), 1 liter (1 qt)
1	green color-coded stock container of talc (baby powder), 1 liter (1 qt)
1	blue color-coded stock container of baking soda, 1 liter (1 qt)
1	orange color-coded stock container of cornstarch, 1 liter (1 qt)

Testing Liquids

2	bottles of white vinegar, 250 ml (½ pt)
2	bottles of 0.1% iodine solution, 250 ml (½ pt)
2	bottles of red cabbage juice, 250 ml (½ pt)

Household Chemicals for Acids and Bases Test

1	bottle of white vinegar, 250 ml (½ pt)
1	bottle of lemon juice, 250 ml (½ pt)
1	bottle of 2% ammonia solution, 250 ml (½ pt)
1	bottle of 2% rubbing alcohol solution, 250 ml (½ pt)
1	bottle of 2% detergent solution, 250 ml (½ pt)

*	Science notebooks
*3	rolls of paper towels
*2	sponges
*1	whisk broom and dustpan
*1	plastic-lined trash can
*1	dishpan or bucket
*	Newspaper
*2	rolls of masking tape
*1	match jar

*30 student scissors
  * Glue
*1 box of thumbtacks
  * Newsprint, 61 x 91 cm, 24 x 36"
*4 sheets of poster board
*1 plastic trash bag
  * Gallon container
  * Assorted common school objects (see pg. 8)
**31 pairs of safety goggles

Note: These items are not included in the kit. They are commonly available in most schools or can be brought from home.

**We strongly recommend that you and your students wear safety goggles during the unit. Though not included in this kit, the goggles are available from Carolina Biological Supply Company, Catalog No. 97-2160. To order, call toll free 1-800-334-5551.

Teaching *Chemical Tests*

The following information on unit structure, teaching strategies, materials, and assessment will help you give students the guidance they need to make the most of their hands-on experiences with this unit.

Unit Structure

How Lessons Are Organized in the Teacher's Guide: Each lesson in the *Chemical Tests* Teacher's Guide provides you with a brief overview, lesson objectives, key background information, materials list, advance preparation instructions, step-by-step procedures, and helpful management tips. Many of the lessons have recommended guidelines for assessment. Lessons also frequently indicate opportunities for curriculum integration. Look for the following icons that highlight extension ideas for math, reading, writing, oral presentations, art, and social studies.

Please note that all record sheets and blackline masters may be copied and used in conjunction with the teaching of this unit.

Student Activity Book: The *Chemical Tests* Student Activity Book accompanies the Teacher's Guide. Written specifically for students, this activity book contains simple instructions and illustrations to help students understand how to conduct the activities in this unit. The Student Activity Book also will help students follow along with you as you guide each lesson, and it will provide guidance for students who may miss a lesson (or who do not immediately grasp certain

activities or concepts). In addition to previewing each lesson in the Teacher's Guide, you may find it helpful to preview the accompanying lesson in the Student Activity Book.

The lessons in the Student Activity Book are divided into the following sections, paralleling the Teacher's Guide:

- **Think and Wonder** sketches for students a general picture of the ideas and activities of the lesson described in the **Overview and Objectives** section of the Teacher's Guide

- **Materials** lists the materials students and their partners or teammates will be using

- **Find Out for Yourself** flows in tandem with the steps in the **Procedure** section of the Teacher's Guide and briefly and simply walks students through the lesson's activities

- **Ideas to Explore,** which frequently echoes the **Extensions** section in the Teacher's Guide, gives students additional activities to try out or ideas to think about

Teaching Strategies

Classroom Discussion: Class discussions, effectively led by the teacher, are important vehicles for science learning. Research shows that the way questions are asked, as well as the time allowed for responses, can contribute to the quality of the discussion.

When you ask questions, think about what you want to achieve in the ensuing discussion. For example, open-ended questions, for which there is no one right answer, will encourage students to give creative and thoughtful answers. You can use other types of questions to encourage students to see specific relationships and contrasts or to help them summarize and draw conclusions. It is good practice to mix these questions. It also is good practice always to give students "wait time" to answer; this will encourage broader participation and more thoughtful answers. You will want to monitor responses, looking for additional situations that invite students to formulate hypotheses, make generalizations, and explain how they arrived at a conclusion.

Brainstorming: Brainstorming is a whole-class exercise in which students contribute their thoughts about a particular idea or problem. When used to introduce a new science topic, it can be a stimulating and productive exercise. It also is a useful and efficient way for the teacher to find out what students know and think about a topic. As students learn the rules for brainstorming, they will become more and more adept in their participation.

To begin a brainstorming session, define for students the topics about which they will share ideas. Tell students the following rules:

■ Accept all ideas without judgment.

■ Do not criticize or make unnecessary comments about the contributions of others.

■ Try to connect your ideas to the ideas of others.

Cooperative Learning Groups: One of the best ways to teach hands-on science is to arrange students in small groups. Materials and procedures for many lessons in *Chemical Tests* are based on groups of four. There are several advantages to this organization. It provides a small forum for students to express their ideas and get feedback. It also offers students a chance to learn from one another by sharing ideas, discoveries, and skills. With coaching, students can develop important interpersonal skills that will serve them well in all aspects of life. As students work, they will often find it productive to talk about what they are doing, resulting in a steady hum of conversation. If you or others in the school are accustomed to a quiet room, this new, busy atmosphere may require some adjustment.

Learning Centers: You can give supplemental science materials a permanent home in the classroom in a spot designated as the learning center. Students can use the center in a number of ways: as an "on your own" project center, as an observation post, as a trade-book reading nook, or simply as a place to spend unscheduled time when assignments are done.

In order to keep interest in the center high, change the learning center or add to it often. Here are a few suggestions of items to include.

■ Science trade books on chemistry, chemicals, and scientists (see the **Bibliography, Appendix B,** for trade-book annotations).

■ Magnifying lenses and an assortment of interesting objects to observe, such as leaves, seeds, salt, soil, rocks, newspaper, fabric, and feathers.

■ Mystery books and books about people who apply problem-solving techniques in their work.

■ Items contributed by students for sharing, such as crystals and favorite subject-related trade books.

Materials

Safety Notes: This unit does not contain anything of a highly toxic nature, particularly given the dilutions and quantities involved, but common sense dictates that nothing be put in the mouth. In fact, it is good practice to tell your students that, in science, materials are never tasted. Students may also need to be reminded that certain items, such as toothpicks, forceps, and plastic droppers, are not toys and should be used only as directed.

You will find that many lessons emphasize good safety practices. In fact, to foster safety awareness in your classroom, you must make it a point to discuss safety rules and, in particular, the wearing of goggles. It is very important that you wear goggles whenever the student are told to do so. Safety goggles are not included in this kit; however, we strongly recommend their use. See the Materials List, pg. 6, for ordering information. Some suggested safety rules to include on your class "Safety Rules" poster are:

■ Do not run, push, or fool around.

■ Do not taste any chemicals in science class.

■ Clean up all spills immediately.

■ Wear goggles when working in science class.

■ Follow directions carefully.

The powders and crystals used in this unit are not hazardous. However, caution students to handle the chemicals according to the directions. This will prevent students from creating excessive airborne particles that could be inhaled.

Additional information about some of the materials you will use appears in **Safety Notes** throughout the unit.

Appendix D explains how to make the red cabbage juice for Lesson 9 and the iodine, ammonia, alcohol, and detergent solutions for Lessons 8 and 15.

Organization of Materials: To help ensure an orderly progression through the unit, you will need to establish a system for storing and distributing materials. Being prepared is the key to success. Here are a few suggestions.

- Know which activity is scheduled and which materials will be used.

- Familiarize yourself with the materials as soon as possible. Label everything and put on new labels if the old ones become unreadable.

- Organize your students so that they are involved in distributing and returning materials. If you have an existing network of cooperative groups, delegate the responsibility to one member of each group.

- Organize a distribution center and train your students to pick up and return supplies to that area. (The most frequent tasks include distributing the science pails and occasionally refilling the bottles of test liquids.) A cafeteria-style approach works especially well when there are large numbers of items to distribute.

- Look at each lesson ahead of time. Some have specific suggestions for handling materials needed that day.

- Streamline cleanup by providing the class with a cleanup box and a packet of paper towels. Students can put disposable materials into this box and clean off their tables at the end of each lesson.

Additional management tips are provided throughout the unit. Look for the icon at the right.

Using Tables and Lists: In this unit, the students learn the value of using tables to record observations and organize results. These tables provide a tangible account of your students' learning processes and experiences. The lists below are introduced at different points. Displaying them in the classroom from lesson to lesson will allow you to add to them and use them as information resources. Some teachers have strung clothesline across the classroom on which to keep as many lists hanging at a time as possible. It is especially useful to keep the following lists on display throughout the unit:

- "What We Think about Chemicals"

- "What We Would Like to Know about Chemicals"

- "Check Your Science Pail" poster

- "Safety Rules" poster

- "How We Are Finding Out about the Unknowns"

Assessment

Philosophy: In the Science and Technology for Children program, assessment is an ongoing, integral part of instruction. Because assessment emerges naturally from the activities in the lessons, students are assessed in the same manner in which they are taught. They may, for example, perform experiments, record their observations, or make oral presentations. Such performance-based assessments permit the examination of processes as well as of products, emphasizing what students know and can do.

The goals for learning in STC units include a number of different science concepts, skills, and attitudes; therefore, a number of different strategies for performance assessment are provided to help you assess and document your students' progress toward the goals. These strategies also will help you report to parents and appraise your own teaching. In addition, the assessments will enable your students to view their own progress, reflect on their learning, and formulate further questions for investigation and research. Figure T-1 summarizes the learning goals for this unit and where they are addressed and assessed.

Assessment Strategies: The assessment strategies in STC units fall into three categories: matched pre- and post-unit assessments, embedded assessments, and final assessments.

The first lesson of each STC unit is a *pre-unit assessment* designed to give you information about what the whole class and individual students already know about the unit's topic and what they want to find out. It often includes a brainstorming session during which students share their thoughts about the topic through exploring one or two basic questions. In the *post-unit assessment* following the final lesson, the class revisits the pre-unit assessment questions, giving you two sets of comparable data that indicate students' growth in knowledge and skills (see Figure T-2).

continued on pg. 12

Chemical Tests: Goals and Assessment Strategies

Concepts	
Goals	**Assessment Strategies**
Common household chemicals have different physical and chemical properties. 　Lessons 1–16	Lessons 1, 3, 7, 9, 11, 12, 15 　▪ Class and team discussions 　▪ Notebook entries 　▪ Record sheets
Chemicals undergo changes in form, color, and texture when they are mixed together, separated, or heated. 　Lessons 1, 3–10	Lessons 4, 6, 9 　▪ Notebook entries 　▪ Record sheets 　▪ Class and team discussions
Some chemicals can be identified by their interaction with water, vinegar, iodine, red cabbage juice, and heat. 　Lessons 11, 12–14, 16	Lessons 11, 12–14, 16 　▪ Record sheets 　▪ Class discussions 　▪ Notebook entries 　▪ Oral presentations
Different types of mixtures, such as solutions and suspensions, are created when solids are combined with water. 　Lessons 4–6, 13–14, 16	Lessons 4, 6 　▪ Class discussions 　▪ Notebook entries 　▪ Record sheets
Evaporation and filtration are methods for separating mixtures of solids and liquids. 　Lessons 5–6	Lesson 6 　▪ Record sheets 　▪ Class lists and discussion
Some chemicals can be classified as acids, bases, or neutral substances by their reactions with red cabbage juice. 　Lesson 15	Lesson 15 　▪ Class lists and discussion 　▪ Notebook entries

Skills	
Goals	**Assessment Strategies**
Observing and describing properties of materials. 　Lessons 1–10, 13–16	Lessons 1, 3–4, 6, 9, 14, 16 　▪ Record sheets 　▪ Class lists and discussions 　▪ Notebook entries
Learning to perform different physical and chemical tests. 　Lessons 3–10, 14	Lessons 3, 9, 14, 16 　▪ Record sheets 　▪ Notebook entries 　▪ Observation of testing procedures
Predicting, observing, describing, and recording results of tests. 　Lessons 3–10, 13–16	Lessons 1, 4, 6, 9, 14, 16 　▪ Notebook entries 　▪ Class discussions 　▪ Record sheets
Analyzing and drawing conclusions from the results of tests. 　Lessons 4–16	Lessons 6, 9, 11–12, 14–16 　▪ Notebook entries 　▪ Record sheets 　▪ Class discussions

Goals	Assessment Strategies
Comparing and contrasting test results to define the properties of household chemicals so they can be identified. Lessons 11–16	Lessons 11, 14, 16 • Notebook entries • Record sheets • Class lists and discussions • Teacher observations
Supporting conclusions with reasons based on experiences. Lessons 6, 12–16	Post-unit assessments and Lessons 6, 9, 12–16 • Notebook entries • Record sheets • Class lists and discussions
Communicating results and reflecting on experiences through writing and discussion. Lessons 1–16	Lessons 4, 6, 9, 11–12, 14, 16 • Notebook entries • Class and team discussions • Teacher's observations • Oral presentations
Applying previously learned knowledge and skills to new situations to solve a problem. Lessons 13, 14–16	Lessons 14–16 • Notebook entries • Class discussions • Record sheets • Teacher's observations
Reading to enhance understanding of chemistry concepts. Lessons 6, 12, 15	Lessons 6, 12, 15 • Class discussions • Teacher's observations
Developing proper lab techniques to ensure safety and avoid contamination. Lessons 2-10, 13–16	Lessons 3, 9, 16 • Teacher's observations

Attitudes	
Goals	**Assessment Strategies**
Developing interest in and enthusiasm toward exploring and investigating properties of chemicals. Lessons 1–16	Lessons 1, 16, Appendix A • Teacher's observations • Student self-assessment
Recognizing the importance of guidelines for experimentation. Lessons 3–10, 13–16	Lessons 3, 9, 16, Appendix A • Teacher's observations • Student self-assessment
Developing an awareness of the importance of chemicals in our lives. Lessons 1, 12, 15	Lessons 1, 9, 12, Appendix A • Pre- and post-unit assessments • Notebook entries • Student self-assessment
Developing an appreciation for the safe handling of chemicals. Lessons 1–16	Lessons 1–3, 9, 16 • Teacher's observations • Class discussions • Class list

Figure T-2

Sample of matched pre- and post-unit class assessments

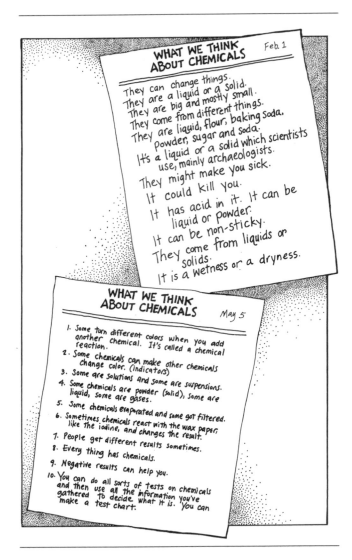

continued from pg. 9

Throughout a unit, assessments are woven into, or embedded, within lessons. The activities used as embedded assessments are indistinguishable from those in lessons. For embedded assessments, however, the teacher records information about students' learning. Whatever the assessment activity, all are intended to provide an ongoing, detailed profile of students' progress and thinking.

Opportunities for embedded assessments occur at natural points in a unit. In many STC units, as in Chemical Tests, the last lesson is also an assessment activity that challenges students to synthesize and apply much that they have encountered in the previous lessons. Specific guidelines for a variety of assessments are found at the end of several lessons.

Appendix A contains additional final assessments that can be used to document students' understanding after the unit has been completed. In these assessments, students may solve problems through the hands-on application of materials or the interpretation and organization of data. Students may also plan and carry out an experiment. On occasion, an appropriate paper-and-pencil test is included. In addition, **Appendix A** includes a self-assessment that helps students reflect on their learning. When you are selecting final assessments, consider using more than one assessment to give students with different learning styles additional opportunities to express their knowledge and skills.

Documenting Student Performance: In STC units, assessment is based on your recorded observations and students' work products and oral communication. All these documentation methods together will give you a comprehensive picture of each student's growth.

Teachers' *observations and anecdotal notes* often provide the most useful information about students' understanding. Because it is important to document observations used for assessment, teachers frequently keep note cards, journals, or checklists. Many lessons include guidelines to help you focus your observations. Each day, you should try to record your observations of a small group of students. By the end of the unit, you will have numerous observations for every student in

your class. The blackline master on pg. 14 provides a format you may want to use or adapt for recording observations. It includes this unit's goals for science concepts and skills.

Work products, which include both what students write and what they make, indicate students' progress toward the goals of the unit. Examine students' work regularly; their written materials should be kept together in their science notebooks to document learning over the course of the unit. When students refer back to their work from previous lessons, they can reflect on their learning.

A variety of materials are produced during a unit. Record sheets—for example, written observations, drawings, graphs, tables, and charts—are an important part of all STC units. They provide evidence of each student's ability to collect, record, and process information. Students' science notebooks or journals are another type of work product. Often a rich source of information for assessment, notebooks reveal how students have organized their data and what their thoughts, ideas, and questions have been over time. In some cases, students do not write or draw well enough for their products to be used for assessment purposes, but their experiences in trying to express themselves on paper are nonetheless beneficial. Other work products might include Venn diagrams, posters, and written research reports.

Oral communication—what students say formally and informally in class and in individual sessions with you—is a particularly useful way to learn what students know. Ongoing records of class and small-group discussions should be a part of your documentation of students' learning.

Interviews with your students can be used both to explore their thoughts and to diagnose their needs; patterns in students' thinking often surface, for example, when carefully formulated questions stimulate students to explain their reasoning or the steps they used in a process. The questions that students themselves ask also can be a valuable source of information about their understanding. Individual and group presentations can give you insights about the meanings your students have assigned to procedures and concepts and about their confidence in their learning; in fact, a student's verbal description of a chart, experiment, or graph is frequently more useful for assessment than the product or results. Questions posed by other students following presentations provide yet another opportunity for you to gather information.

Blackline Master
Chemical Tests: Observations of Student Performance

STUDENT'S NAME:	

Concepts	**Observations**
• Common household chemicals have different physical and chemical properties.	
• Chemicals undergo changes in form, color, or texture when they are mixed together, separated, or heated.	
• Some chemicals can be identified by their interaction with water, vinegar, iodine, red cabbage juice, and heat.	
• Different types of mixtures, such as solutions or suspensions, are created when solids are combined with water.	
• Evaporation and filtration are methods for separating mixtures of solids and liquids.	
• Some chemicals can be classified as acids, bases, or neutral substances by their reactions with red cabbage juice.	

Skills	
• Observing and describing properties of materials.	
• Learning to perform different physical and chemical tests.	
• Predicting, observing, describing, and recording results of tests.	
• Analyzing and drawing conclusions from the results of tests.	
• Comparing and contrasting test results to define the properties of household chemicals so they can be identified.	
• Supporting conclusions with reasons based on experiences.	
• Communicating results and reflecting on experiences through writing and discussion.	
• Applying previously learned knowledge and skills to new situations to solve a problem.	
• Reading to enhance understanding of chemistry concepts.	
• Developing proper lab techniques to ensure safety and avoid contamination.	

Thinking about Chemicals

Overview and Objectives

This introductory lesson will provide you with a pre-unit assessment of your students' current knowledge of chemicals and questions about them. Through class discussion and their observations of a common chemical, students will provide you with information you can use to assess both their initial ideas about chemicals and their observation and recording skills. This lesson introduces students to the concept of properties and sets the stage for the activity in the next lesson: observing the properties of common classroom objects.

■ Students set up a notebook to record their observations and ideas.

■ Students share their present thinking about chemicals and discuss what they would like to learn about them.

■ Students gain experience observing an unknown material and describing its properties.

Background

Chemistry is the study of chemicals and how they interact. Chemicals consist of the 106 known elements and any of their combinations (the chemical water, for example, consists of the elements hydrogen and oxygen). Everything is a chemical, whether it is made by nature or by humans. Our bodies are made of chemicals, and we produce chemicals such as carbon dioxide. We need chemicals from food and the air we breathe.

In this first lesson, students observe and describe an unknown or "mystery" chemical. This chemical—cornstarch—is one of the five unknowns the class will investigate in Lessons 2 through 11. (While students will not know the chemical's identity at this time, they will apply their knowledge and skills to discover it in Lesson 13.) This activity introduces the class to the concept of **properties,** those characteristics used to describe an object. Here, students use their senses to examine physical properties such as color, shape, texture, and odor. To ensure safety, they do not use their sense of taste at any time in this unit.

At this time, students also produce class lists of what they now think about chemicals and what they would like to learn (see Figure 1-1) as well as individual notebook entries that reveal some of their current thinking on chemicals. Both products are used for pre- and post-unit assessment. (See **Post-Unit Assessment** on pg. 171 for more information.)

Note: Many teachers find that their students have had little experience observing and describing the properties of objects. Before you begin the unit, have students try one or more of the observing and describing activities in **Appendix E.**

Figure 1-1

Sample pre-unit
assessment charts

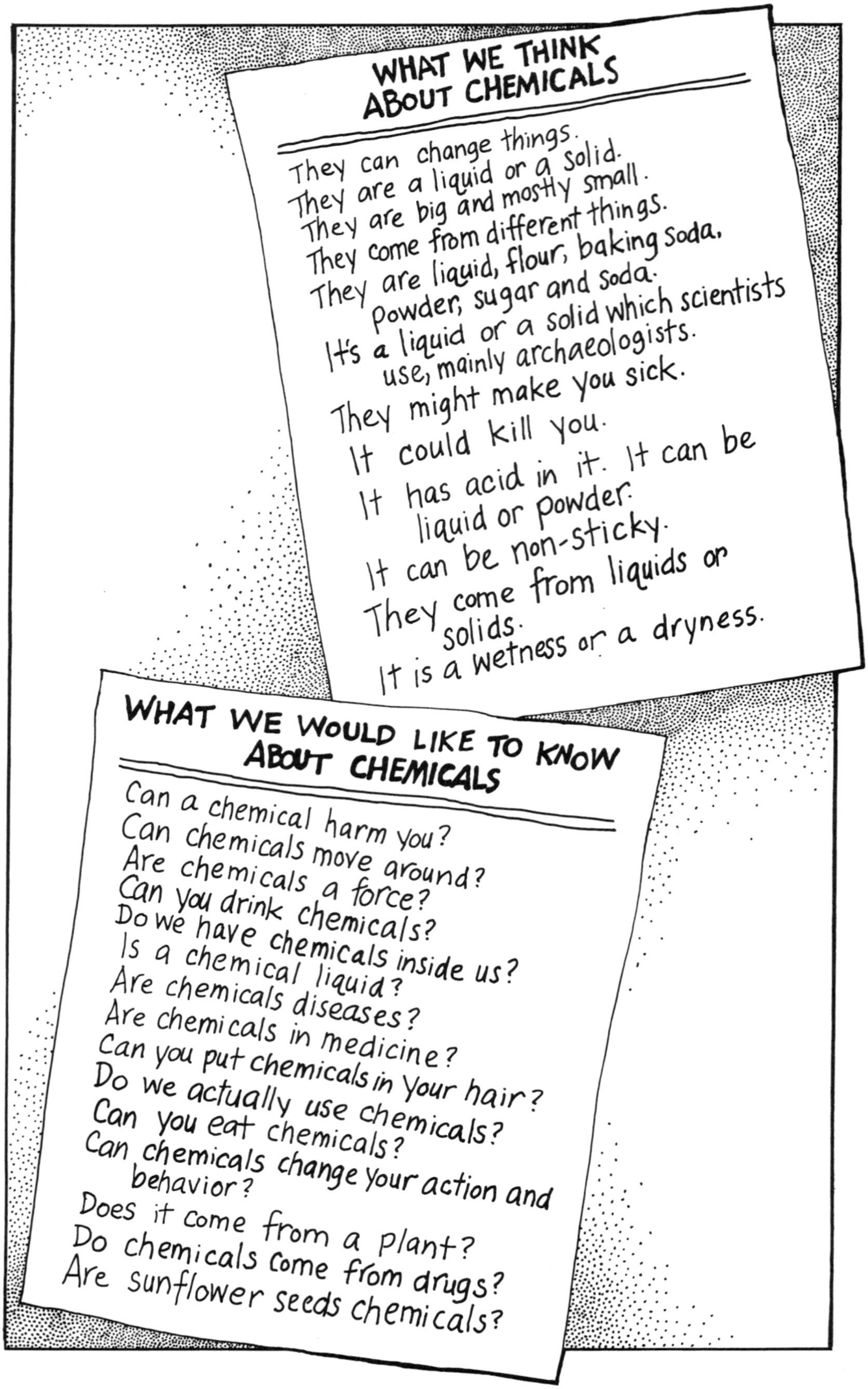

WHAT WE THINK ABOUT CHEMICALS

They can change things.
They are a liquid or a solid.
They are big and mostly small.
They come from different things.
They are liquid, flour, baking soda, powder, sugar and soda.
It's a liquid or a solid which scientists use, mainly archaeologists.
They might make you sick.
It could kill you.
It has acid in it. It can be liquid or powder.
It can be non-sticky.
They come from liquids or solids.
It is a wetness or a dryness.

WHAT WE WOULD LIKE TO KNOW ABOUT CHEMICALS

Can a chemical harm you?
Can chemicals move around?
Are chemicals a force?
Can you drink chemicals?
Do we have chemicals inside us?
Is a chemical liquid?
Are chemicals diseases?
Are chemicals in medicine?
Can you put chemicals in your hair?
Do we actually use chemicals?
Can you eat chemicals?
Can chemicals change your action and behavior?
Does it come from a plant?
Do chemicals come from drugs?
Are sunflower seeds chemicals?

Materials

For each student

1 science notebook, either looseleaf or with pockets (to insert record sheets)
1 pencil
1 pair of goggles (not included in kit; see Materials List, pg. 6)
1 resealable plastic bag, 1 liter (1 qt)

For every two students

1 *Chemical Tests* Student Activity Book

For every four students

1 plastic cup, 207 ml (7 oz), containing approx. 28 ml (1 oz) of water
1 resealable plastic bag, 1 liter (1 qt), containing mystery chemical (cornstarch), 60 cc (¼ c)
2 paper towels
1 tray

For the class

Several sheets of newsprint or poster board and a large marker
1 measuring cup, 60 cc (¼ c)
1 plastic graduated cylinder, 50 ml (1¾ oz)
2 examples of chemicals: cup of water and one plastic mystery bag of unknown chemical (cornstarch)
1 plastic dropper
Cleanup supplies (dustpan and broom, plastic-lined trash can, sponges or whisk broom)
1 plastic trash bag
2 rolls of masking tape

Preparation

1. Label two sheets of newsprint or poster board "What We Think about Chemicals" and "What We Would Like to Know about Chemicals," respectively. You may need to add extra sheets.

2. Set up a materials center in the room (see Figure 1-2). Arrange the materials "cafeteria style" so students can pick up everything they need.

    ■ Select one large area (or several small areas) of the room where students can easily walk by in single file and on both sides of the supplies.

    ■ Place all the materials in a line on one table or on several desks or tables pushed together.

    ■ Place a printed label in front of each group of items that tells what the items are and how many to take.

3. Using the measuring cup, put ¼ cup of cornstarch into each of 10 heavy-duty plastic bags and seal them.

4. To create the "goo" students will observe in this lesson, you need to add water to the cornstarch. To determine the right amount of water, first add 28 ml of water from the plastic graduated cylinder to one bag of cornstarch. Seal the bag and gently knead it. The goo should appear liquid but feel solid. If the mixture does not have the right consistency, try again with a

Figure 1-2

Materials center

new bag of cornstarch. (The humidity level in your classroom may affect the goo's consistency.) Use a plastic dropper to adjust the amount of water in the graduated cylinder.

Note: If you use more than two bags of cornstarch, be sure to replace them. You will need eight bags for the lesson.

5. Use the graduated cylinder to fill each plastic cup with the appropriate amount of water, as determined in Step 4. Cover them with a trash bag to hide them from the students' view and to reduce evaporation.

6. Have at hand one mystery bag of cornstarch and one cup of water.

7. Arrange the students so they can work easily in groups of four.

8. Read through Lesson 1 as it is presented in the Student Activity Book and decide when in the lesson you want to distribute the books to students.

Management Tip: You may want to teach this lesson in two parts. If so, check the **Procedure** section for a suggested stopping point.

Procedure

1. Explain to your students that for the next eight weeks they will be observing and testing different chemicals. But first, they will share what they already know about chemicals.

2. Have each student write today's date at the top of the first page in his or her notebook. Hold up the cup of water and the plastic bag of cornstarch. Keeping their identities a mystery, explain that both are examples of chemicals. Ask students to think about the following questions:

    ■ What do you know about chemicals?

    ■ Where have you seen or what have you heard about chemicals?

- What are some uses of chemicals?

- How have you used chemicals?

3. Have students share their thoughts with each other in groups of four. Explain that after the discussion one person from each group will report the group's thoughts to the class.

4. Display the sheet "What We Think about Chemicals." As each group representative makes his or her report, record the responses. Put a check next to duplicates to acknowledge all students' ideas.

5. Now tell students that in groups they will observe a common chemical, but they won't know what the chemical is. Focus attention on the materials center. Demonstrate how to walk carefully, take turns, and read the labels. Have one member from each team pick up the following materials:

- 1 tray

- 1 mystery bag of an unknown chemical

- 2 paper towels

- 4 pairs of goggles

- 4 plastic bags for the goggles

6. Explain briefly that when people work with chemicals, it is important always to wear goggles. Hand out the two rolls of masking tape, and have students each take a piece. Ask them to write their names on the tape and to stick it on the plastic goggle bag. Explain that they will store the goggles in these bags throughout this unit. Put on your goggles and have the students do so as well. Show them how to stretch the band to adjust the fit, if necessary.

   Note: Remember to wear goggles yourself whenever your students need to. Also, if students detect a "plastic" smell from the goggles, have them wash the goggles with warm water and soap before wearing them.

7. Remind students that the mystery chemical inside the bag is a common one. Challenge them to discover as much as possible about the mystery chemical without opening the bag. (Typically, students will feel the bag and try to smell it.) Warn the class not to press too hard, or the bags will break. Give students a few minutes to discuss the chemical with their group.

8. Have one student from each group share with the class his or her observations and the senses used to observe the chemicals. Now ask the students how they might find out more about the mystery bag's contents. Most students will suggest opening the bag and touching or smelling the chemical. If not, suggest it yourself, and have students open the bags. When students are ready to smell the substance, demonstrate how to do so (see Figure 1-3).

   Safety Note: When touching the material, students should put a small amount between their fingers and observe it over the tray or on it. Help students understand never to touch an unknown chemical unless they are told it is safe. Ask students to use paper towels to wipe their hands.

9. Help the students summarize what they have discovered so far with questions like the following:

- What senses have you been using in your observations?

- What words would you use to describe the mystery chemical?

   Introduce the term **property** as a word that is used to describe how an object looks, feels, smells, sounds, and so on.

Figure 1-3

Using your sense of smell

10. Ask students how they can reveal even more properties of the unknown chemical. How can they go beyond using their senses alone? If no one suggests adding something to the chemical, ask what might happen if they added water to the bag. Then distribute the cups of water.

11. Modeling the process with your own water and plastic bag, tell students to do the following:

   ■ Slowly pour the water into the bag and seal it.

   ■ Carefully knead the bag, evenly mixing the water with the chemical to create goo.

12. Ask the following questions and have students share their new observations.

   ■ What new properties of the chemical did you discover?

   ■ What do you think caused the changes in the chemical?

   ■ Did adding the water give you more information about the mystery chemical?

   Let students know they will continue to investigate the properties of this chemical as well as others throughout the unit. In this way, they will discover the mystery chemical's identity.

   Note: Do **not** tell students that the mystery chemical they have observed in Lesson 1 is one of the five unknown chemicals they will investigate in Lesson 2.

13. Have the class clean up by returning the trays, cups, and goggles (placed in their labeled plastic bags) to the materials center. Ask students to throw away the sealed bags of goo and the paper towels and to wipe desks clean.

Management Tip: If you have decided to teach this lesson in two parts, this is one possible stopping point.

Final Activities

1. Display the sheet "What We Would Like to Know about Chemicals." Ask students to share their thoughts. Record them on the sheet. Leave this list, as well as "What We Think about Chemicals," posted in the classroom.

2. Have students respond to the following scenario in their notebooks:

   ■ A label has peeled off a container of white powder in your kitchen. Describe some ways you could find out what the powder is.

   You and your class will revisit this pre-unit activity at the end of the unit (see **Post-Unit Assessment,** pg. 171).

Figure 1-4

Sample notebook
entry for pre-unit
assessment

Date April 15, 1992

I think it is blue?

→ I would add water。See if it would stick to something.

Date

Shawn
Thomas

3. Have students begin a science journal in their notebooks. Try idea starters such as "I was surprised when _____" or "Today I discovered that _____."

Extensions

1. Ask students to write in their notebooks any ideas about what the mystery chemical could be and why they think so.

2. Create a "Chemical Tests" bulletin board. As you proceed through the unit, note on it descriptive words or phrases connected with the four senses. This will provide a vocabulary board to which students may refer when recording observations (see Figure 1–5).

Figure 1-5

Sample vocabulary bulletin board

Sample Vocabulary Bulletin Board

Sight	Hearing	Touch	Smell
bubbly	crunches	smooth	earth-like

Assessment

This lesson has many assessment opportunities that you can use to establish a base-line for measuring students' progress. You may want to use one or several of them. Record informal observations on charts, cards, or notes to help you keep track of information about each student. Teachers have found it useful to note the following:

- Students' **previous knowledge** of chemicals.

  - What information do students already have?

  - Are students aware that everyday materials are made of chemicals?

  - Are students aware that different chemicals have different properties?

  - Have students had previous experience working with chemicals in school?

  - Is the information they share accurate?

- Students' **observation** skills.

  - Are students able to make observations using all their senses (except taste)?

  - Are students making descriptive observations? ("It's white.") Comparative ones? ("It looks like snow.") Inferences? ("It's white. It must be salt.")?

- Students' **communication** skills.

  - Can students describe objects clearly?

  - Can students describe ideas when working in small and large groups?

Throughout the unit, students will be learning important skills basic to science: observing, recording, and comparing information. You can assess students' progress in these areas in two ways: (1) by observing and talking to students as they work individually and in groups; and (2) by taking a look at individual student products. Both approaches are important.

In the section **Teaching Chemical Tests,** on pgs. 9 to 14, you will find a detailed discussion about the assessment of students' learning. The specific goals and related assessments for this unit are summarized in Figure T-1 on pg. 10. Please keep in mind that some third-graders may not completely understand every concept listed or master every skill. As you observe your class, look for the development of these ideas and skills in your students rather than their mastery.

Figure 1-6

Sample notebook entry for pre-unit assessment

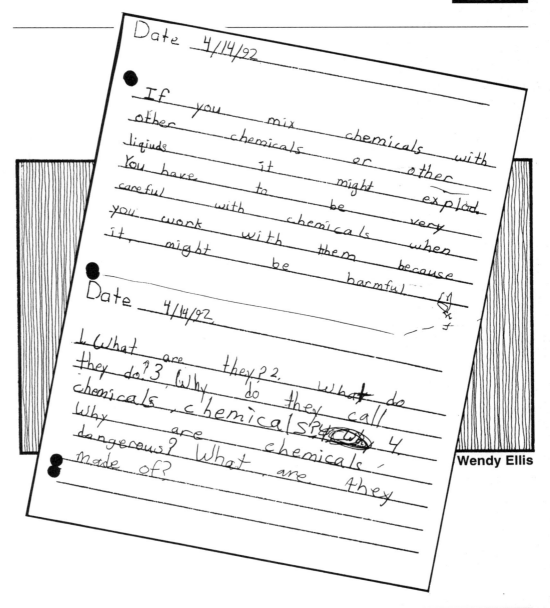

Date ___4/14/92___

• If you mix chemicals with other chemicals or other liqiuds it might explod. You have to be very careful with chemicals when you work with them because it might be harmful.

Date ___4/14/92___

1. What are they? 2. What do they do? 3. Why do they call chemicals, chemicals? 4. Why are chemicals dangerous? What are they made of?

Wendy Ellis

Management Tip: Before the next lesson, assemble the "Check Your Science Pail" poster (see **Preparation** section of Lesson 2 and **Appendix C**).

Note: Before beginning Lesson 2, have your students collect some common objects from school and home for you to place in the mystery object bags (see **Preparation** section of Lesson 2). The objects (such as a scissors, pencil, marker, ruler, spoon, or rubber band) should be familiar to students.

Investigating Unknown Solids: Getting Ready

Overview and Objectives

As a way to build on the concept of properties introduced in Lesson 1, students observe and describe the properties of classroom objects. This experience will improve students' ability to observe the properties of the five unknowns (which are common household chemicals) that they will use in Lesson 3. Students also are introduced to the tools they will use to conduct their investigations of these chemicals.

- Students assemble the tools and five unknown chemicals they will be investigating.

- Students learn about the importance of safety in science class.

- Students observe and describe the properties of common classroom objects.

Background

In this unit, the five solid chemical unknowns are color-coded to enable you and your students to distinguish easily among them. You can refer to each as the "blue unknown," the "red unknown," and so on. Students will not discover the identities of the five chemical unknowns until Lesson 12. For your own information, however, they are revealed in Figure 2-1. To sustain interest and make the investigation exciting for your students, it is essential that you do not reveal the identities of the unknowns until Lesson 12.

Figure 2-1

Identities of the Five Unknowns Table

Unknown's Color Code	Unknown
Red	Sugar
Yellow	Alum
Green	Talc (baby powder)
Blue	Baking soda
Orange	Cornstarch

In preparation for this lesson you will need to organize the materials for the entire unit. While this preparation takes more time than usual, it will help save you time in lessons to come. (You may want to ask another adult to assist with this preparation, the lesson itself, or both.)

The first thing you will need to do is set up the following three areas (see Figures 2-2 and 2-3):

■ Materials center

■ Cleanup area

■ Pail storage area for assembled science pails

Figure 2-2

Materials center and pail storage area

The cleanup and pail storage areas can be part of the materials center or separate, depending on your classroom space.

Throughout the unit, students probably will complete activities at different rates. Students who finish early can use their free time to link their experiences to language arts. One way to do this is to read mysteries. (See the **Bibliography** for a list of mystery books appropriate for third-graders.)

Note: Your students will be working in teams of two from now through Lesson 10. (See **Teaching Chemical Tests** on pg. 7 for information on effective team building.)

Materials

For each student
   1   science notebook
   1   pencil

For every two students
   1   science pail
   5   jars, 14 g (½ oz)
   1   plastic bag, 1 liter (1 qt), containing five measuring spoons
   2   pairs of goggles in their bags
   5   colored dots (red, orange, blue, green, yellow), .5 cm (¼")
 10   colored dots (2 red, 2 orange, 2 blue, 2 green, 2 yellow), 2 cm (¾")
   1   brown paper lunch bag containing one common object

Figure 2-3

Cleanup area

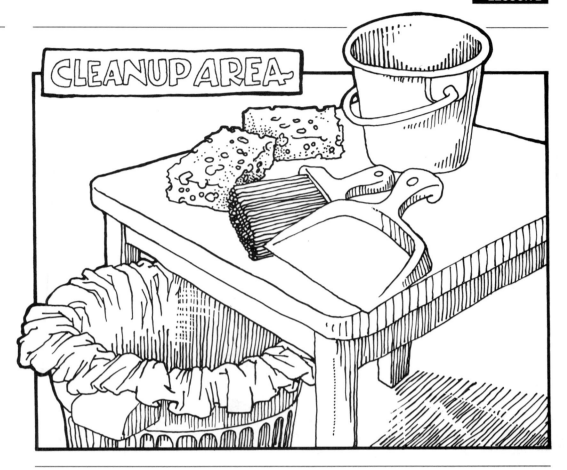

For the class

  5   stock containers of the five unknowns, 1 liter (1 qt)
  5   color-coded jars containing the five unknowns
  5   color-coded measuring spoons
      Newspaper
  1   roll of masking tape
  1   "Check Your Science Pail" poster (see **Appendix C**)
  1   sheet of newsprint or poster board
      Assorted markers
      Cleanup supplies

Safety Note: The talc used in this unit is the common baby powder sold in drugstores. Handling this material should create a minimal—and nonhazardous—amount of talc dust. If inhalation does occur, contact a physician immediately.

Preparation

1. Put a strip of masking tape on each pail. Later, teams will be asked to label the pails with their names.

2. Have student helpers cut apart each sheet of large colored dots into seven sets so that each team easily can pick up two sets of each color. Then have helpers cut apart the sheets of small colored dots so teams can pick up one set of each color. Put the two sizes of dots in separate containers.

3. Stock the materials center with the following:

   ■ 15 trays

   ■ paper towels

   ■ unknown jars

   ■ 15 science pails

   ■ 15 plastic bags, each containing 5 measuring spoons

   ■ 30 labeled bags, each containing a pair of goggles

   ■ 80 small dots of each color (red, orange, yellow, blue, green)

   ■ 160 large dots of each color (red, orange, yellow, blue, green)

4. Set up a jar-filling area large enough to accommodate five teams of students at a time, and cover the space with newspaper. Use small dots to color-code one set of measuring spoons and place them next to the corresponding color-coded stock containers of unknowns. See Figure 2-4 for the materials needed at the jar-filling station.

Figure 2-4

Jar-filling station

 Management Tip: This setup rotates students through a single station to fill their jars with the unknowns. Alternatively, you could set up five different jar-filling stations, one for each unknown.

5. At the jar-filling station, fill one set of five jars with the unknowns to use as a model in Step 1 of the **Procedure.** Color-code each of the jars with two large dots: one on the lid and one on the side.

   Note: If the unknown in the stock container coded with the yellow dot contains clumps, use a spoon or similar object to break them into powder.

6. Assemble and hang the "Check Your Science Pail" poster (see **Appendix C** for instructions). This will help students keep their pails complete and speed along cleanups. Attach the goggles strip, unknowns strip, and measuring spoons strip to the poster.

7. Stock the cleanup area with plastic-lined trash cans, a broom, dustpan, and sponges or a whisk broom.

8. Label the newsprint sheet "Safety Rules" and hang it.

9. Team your students in pairs.

10. Place one of the common objects students collected earlier into each brown paper bag. Have these on hand for the **Final Activities.**

Management Tip: At this time, you may want to plan an extra activity for students to do while they are waiting to fill their jars of unknowns.

Procedure

1. Hold up your set of jars containing the five unknowns and explain that each jar contains a common chemical often found at home. Let students know that over the next few weeks, they will perform experiments to gather information and solve a mystery: What are the identities of the five chemical unknowns?

2. Ask students to suggest some ways they can gather evidence about the unknowns. If they find this difficult, remind them how they collected information about the unknown chemical in Lesson 1.

3. Explain that to solve the mystery, each team will need its own set of five unknowns as well as several tools. So, each team will put together a science pail of tools to use throughout the unit. Goggles are one important tool. Ask students the following questions:

   ■ Why was it important to wear goggles in the first lesson?

   ■ Why is safety important?

   On the "Safety Rules" poster, write: "Wear goggles when working in science class" and "Follow directions carefully." Tell students they will help you add to this list as they progress through the unit. (The **Teaching Chemical Tests** section on pg. 7 suggests additional safety rules.)

4. Have one member from each team pick up materials. Have the team members write their names on the tape on their science pail.

5. Explain the purpose of the jar-filling station, and show the class how to use the large dots to color-code the jars and lids. Have students do it.

6. Then explain that five teams will be at the station at a time, one team at each unknown. Also tell students that while they wait for their turn at the station, they will color-code their measuring spoons by placing a small dot of a different color on the handle of each spoon. Review the following instructions (on pg. 8 of the Student Activity Book):

   ■ Put on your goggles.

   ■ With your partner, bring your science pail (with the five sample jars in it) to the jar-filling station.

   ■ Take out the jar whose colored dot matches the colored dot on the container at your work space.

   ■ Unscrew the lid of your jar. Take the lid off the container at your space. Using the spoon for that container, carefully fill your jar.

   ■ When you have finished filling your jar, be careful to screw the lid on tightly. Put the jar back in your science pail. Put the lid back on the container at your workspace.

Figure 2-5

Filling the unknown sample jars

- Wait until the other teams have finished filling their jars. With your partner, move to the next workspace to your right. The other teams will move, too.

- Repeat the same steps. Be sure the colored dot on the jar you are filling matches the dot on the container at your new workspace.

- When you have filled all five jars, bring your pail back to your regular seat.

- Now finish color-coding your measuring spoons.

- After you've filled your jars and color-coded your measuring spoons, look at Figure 2-6. That's what you should have at the end of this activity.

Figure 2-6

Fully assembled pail

7. Students will need to clean up after each lesson. You can help cleanup proceed smoothly by doing the following:

   ■ Encourage every student to do a fair share of the work.

   ■ Make sure each team knows it is responsible for returning all of its materials to the materials center.

   ■ Ask students to look at the "Check Your Science Pail" poster. Explain that it will show what should be in their pails at the end of every lesson.

   ■ Remind students where the pails will be stored and that they are stackable.

Final Activities

Management Tip: You may choose to do the **Final Activities** during language arts.

1. Tell students that before they explore and describe the properties of the chemical unknowns, they will practice observing and describing known objects. Then ask them to take a clean sheet of paper out of their notebooks.

2. Explain the activity: to guess the identity of another team's "mystery object" on the basis of a short description that the team has written. This description must focus only on properties you can see, feel, or smell.

3. Then, write on the chalkboard a "What Am I?" description of an object such as a pencil. For example:

   I am yellow. I have the number 2, Mongol, and FaberCastell written on me in black. I am sharp on one end. I have a red rubber tip on the other end. What am I?

4. Hold up a sample mystery object bag and give the following directions to the teams:

   ■ Open the bag and closely observe your mystery object, but do not take it out of the bag.

   ■ Use properties you observe to write a brief "What Am I?" description of it.

   ■ Do not include pictures or tell how the object is used.

5. Distribute the sealed mystery object bags.

6. After the teams have written their descriptions, pair each team with another. Have each team swap descriptions—not mystery object bags—with its partner team. After each team has read its description, the other team has one best guess at naming the object. The team that has read the description will reveal the object's identity.

7. Let students perform the "What Am I?" activity.

8. After the students have finished, help them sharpen their observation skills by asking them to respond to some key questions, such as these:

   ■ How many teams guessed the mystery objects?

   ■ Why do you think you were able to guess what the object was?

   ■ How could you change the descriptions to help more teams guess the objects?

Extensions

1. Continue to build descriptive language skills by having the students write more "What Am I?" riddles in categories they choose, such as animals or food. Challenge students to write a "What Am I?" riddle using only math language. For example, "I have four corners. I am four inches long."

2. Have students write **cinquain poems** (see Figure 2-7). Explain that these poems describe an object in terms of observable properties (that is, things you can see, hear, smell, or feel). The poems have five lines and follow a specific pattern:

   ■ Noun—person, place, or thing (very general noun)

   ■ Adjective, adjective—two properties of the noun

   ■ Descriptive phrase—three-word phrase describing something the noun does or is used for, or something that happens to the noun

   ■ Adjective, adjective—two additional properties of the noun

   ■ Synonym for the noun—specific name of the object

   Have students read the first lines of their cinquain poems and cover up the last line. Invite the class to guess what the object is.

Figure 2-7

Sample cinquain poem

Matthew Stevens   Date. 3/30/92

Animal
furry, tickles
Climbs in Cage
Small, quiet

hamster

Matthew Stevens

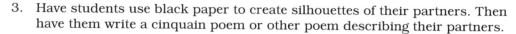

3. Have students use black paper to create silhouettes of their partners. Then have them write a cinquain poem or other poem describing their partners.

4. Post a student's description of a mystery object in the room. Provide a shoebox or an envelope where students can submit their answers.

Management Tip: In Lesson 3, you will assemble students' test mats (see the blackline masters on pgs. 46–47). If possible, laminate these test mats. Otherwise cover them with plastic wrap.

Note: Please follow these instructions at this time. They will help you structure the **Final Activities** for Lesson 5. Do not let your students know you are conducting this experiment.

In Lesson 5, students will set water mixtures aside to evaporate and form crystals, which they will observe in Lesson 6. The rate of evaporation will depend on the conditions in your classroom. To test the evaporation rate for your room, follow these steps:

■ Take a small plastic cup, 37 ml (¼ oz), and an evaporation dish from the kit. Both materials are used in Lesson 5.

■ Fill the cup with water up to the 10-ml line and pour it into the dish.

To speed evaporation and produce smaller crystals, place the dish next to a window (or where the most natural light, air flow, and warmth occur.)

Exploring the Five Unknown Solids

Overview and Objectives

In Lesson 2, students gained experience observing and describing properties of common objects. Now, students are challenged to observe and describe the properties of the five unknown, similar-looking white solids. Students will begin to observe how these chemicals are alike and different in texture, odor, and appearance.

- Students observe the five unknowns using their unaided senses, as well as equipment that extends and enhances their senses.

- Students discuss the properties of the five unknowns.

- Students begin recording and organizing data in a systematic way.

Background

Our senses alone can tell us only so much. To extend our senses, we rely on technology—from simple magnifying lenses to enormously complex microscopes, for example.

There are two kinds of properties: physical and chemical. In this lesson, students explore the **physical properties** of each unknown; that is, properties that can be observed without altering the chemical makeup of the material (e.g., color or shape).

The five unknowns in this unit look similar to the unaided eye (see Figure 3-1 for their identities). When students observe them under a hand lens, however, they quickly discover that the unknowns' physical properties differ. After further investigating physical properties in the next three lessons, students will begin to investigate the unknowns' chemical properties in Lesson 7.

Figure 3-1

Identities of the Five Unknowns Table

Unknown's Color Code	Unknown
Red	Sugar
Yellow	Alum
Green	Talc (baby powder)
Blue	Baking soda
Orange	Cornstarch

As you encourage students to make new discoveries, it may help you to keep in mind some examples of what past children have said:

■ The red unknown looks like tiny crystals.

■ When a small sample of each unknown is rubbed on a piece of black paper, the green, orange, and yellow unknowns leave a mark on the paper; the blue and red unknowns do not.

■ Rubbing the unknowns against the black paper produces certain sounds. Students have said the red unknown makes a "crunchy" sound, the blue unknown makes a "sandy" sound, and the other three unknowns make "soft" sounds.

■ When you manipulate unknowns with a toothpick, some separate into discernible pieces; others do not.

Your students probably will come up with more ways to use their equipment and may suggest properties not mentioned here.

At first, students are likely to offer inferences and opinions rather than scientific observations. At this point, it is important to encourage them to share all thoughts. As students gain more experience observing, their ability to make clear and accurate scientific observations is likely to improve.

 Management Tip: If your students have not used hand lenses before, take a little time to explore their use. An effective way to use the hand lens is to hold the object stationary while keeping the hand lens above it. Then move the lens back and forth until the object is in focus. You may also want to set up a learning center on lenses. Provide a variety of objects (e.g., feathers, rocks, insects, seeds) for the students to observe.

Materials

For each student
    1 science notebook
    1 pencil

For every two students
    1 science pail
    2 hand lenses
    2 pieces of black construction paper, 5 x 11 cm (2 x 4½″)
    1 tray
    1 test mat
    1 sheet of wax paper, approximately 16 x 22 cm (6 x 8½″)
    5 toothpicks
    2 paper towels

For the class
    1 sheet of newsprint
    2 sheets of poster board, 56 x 71 cm (22 x 30″)
    5 colored markers (red, yellow, blue, green, and orange)
       "Safety Rules" poster
       Glue stick or tape
       Cleanup supplies

Preparation

1. Label a newsprint sheet "How We Are Finding Out about the Unknowns." Hang it up.

2. Make the test mats using the blackline masters on pgs. 46–47. (The test mats contain an unlabeled circle, which students will use in Lesson 5.)

3. Have student helpers cut 15 sheets of wax paper, 16 x 22 cm (6 x 8½"), and trim them to fit neatly over the test mat. Cut the black construction paper into 30 strips, approximately 5 x 11 cm (2 x 4½") each.

4. Place the trays, test mats, toothpicks, paper towels and wax paper in the materials center (see Figure 3-2). You will not distribute the black construction paper and the hand lenses until Step 5 of the **Procedure** section.

Figure 3-2

Materials center

5. Add these two strips to the "Check Your Science Pail" poster: hand lenses and black paper.

6. Display the "Safety Rules" poster in an accessible place. Continue to add new rules when needed.

7. Make a "Class Properties Table" by taping together two pieces of poster board and writing the title on it (see Figure 3-3). Divide it according to the different unknowns but not according to the properties; the class will decide on those categories. Have the table on hand for the **Final Activities.**

Procedure

1. For a brief review, ask students the following questions:

    ■ What mystery were you asked to solve in the last lesson?

    ■ What were some of your ideas about how to find out about the unknowns?

2. Explain that the class now will observe the unknowns closely to discover as much as possible about each one. Students will use all senses except taste. On the sheet "How We Are Finding Out about the Unknowns," record the first step in the investigation: observe the unknowns.

3. Now focus attention on the materials center. Review which materials—and how many of each—students will pick up. Have the teams pick up their materials.

Figure 3-3

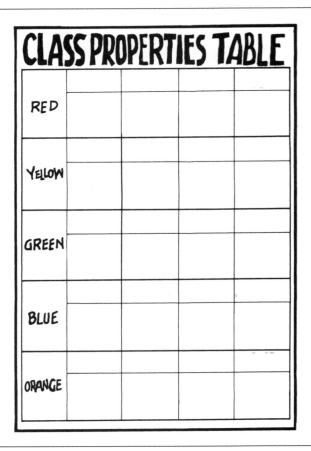

CLASS PROPERTIES TABLE

RED			
YELLOW			
GREEN			
BLUE			
ORANGE			

4. Review the **Student Instructions for Taking a Sample of the Unknowns** on pg. 44 of this guide (pg. 15 of the Student Activity Book). Let the teams begin.

5. After about 10 minutes, distribute two hand lenses and two pieces of black paper to each team. Ask students to briefly share ideas about how they might use these materials to enhance their observations.

6. Have students use the lenses and paper to observe the unknowns once more. Tell them to discuss with their partners the best way to organize their observations. Have students record observations about each unknown in their notebooks.

7. After students have finished recording their observations, have them clean up by following the steps in Figures 3-1 and 3-2 on pg. 12 in the Student Activity Book (Figures 3-4 and 3-5 in this guide).

   ■ Fold the wax paper tightly and throw it, along with the used toothpicks, in the trash.

Figure 3-4

Cleaning up

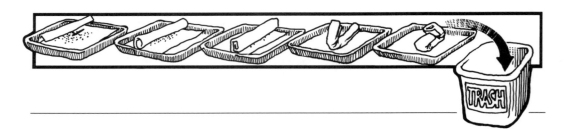

Figure 3-5

Putting materials back in the science pail

- Check your science pail. Remember, the "Check Your Science Pail" poster will show you what should be in your pail.

- Return the other materials to the materials center and put your science pail in the pail storage area. Remember to clean your space.

8. Have teams share the ways they organized their observations of the unknowns. Ask for reasons why they chose each method of organizing. Ask questions such as:

- Why do you think it might be important to record your observations?

- How could organizing your information help you solve the mystery of the five unknowns' identities?

Final Activities

1. Display the "Class Properties Table" you have prepared. Explain that you would like students to help you create a class record of their observations.

2. Point out that you have divided the table into five sections, one for each unknown. Ask students to suggest headings for each column. The suggestions should be based on their observations of the unknowns. Decide as a class which headings to use. (See Figure 3-6 for a sample table organized according to how each sense perceived the unknown. Your students may suggest other ways to organize this table.)

3. Give one student a red marker and ask him or her to record the class's observations for the red unknown on the "Class Properties Table." Repeat this for the other four unknowns. Have students use a matching colored marker for each unknown.

4. To review the concept of properties, ask students to use the "Class Properties Table" to name a property of each of the five unknowns.

Figure 3-6

Sample table

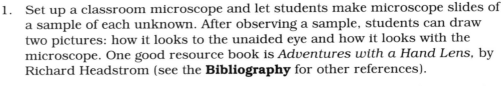

CLASS PROPERTIES TABLE

	HOW IT LOOKS	HOW IT SMELLS	HOW IT FEELS	OTHER OBSERVATIONS
RED				
YELLOW				
GREEN				
BLUE				
ORANGE				

Extensions

1. Set up a classroom microscope and let students make microscope slides of a sample of each unknown. After observing a sample, students can draw two pictures: how it looks to the unaided eye and how it looks with the microscope. One good resource book is *Adventures with a Hand Lens,* by Richard Headstrom (see the **Bibliography** for other references).

2. Have students create **concrete poems,** which combine art and language. A student uses adjectives, nouns, descriptive phrases, and verbs to describe an object and then "draws" these words to form the object's outline. Students also can add color and the object's natural environment to create a scene (see Figure 3-7).

3. A variation on the "What am I?" riddles in Lesson 2 is the "Describer-Guesser" game. One student chooses an object to describe and writes the description, one phrase at a time, for a partner. After each descriptive phrase, the partner tries to guess the object's identity. The writer's goal is to get his or her partner to guess the object in as few phrases as possible.

4. Give students a rock, a leaf, or another object. Ask them to list all the object's properties. You may want to set up a learning center focusing on observing and describing the properties of objects.

5. Use Attribute Blocks™ or Relationshapes™ to further practice identifying and describing properties. Have students sort the blocks on the basis of increasingly specific properties. For example, first sort the blocks with at least one right angle; second, blocks with at least two equal sides; third, blocks that are red; and, finally, small triangles.

Figure 3-7

Sample concrete poems, one by a student

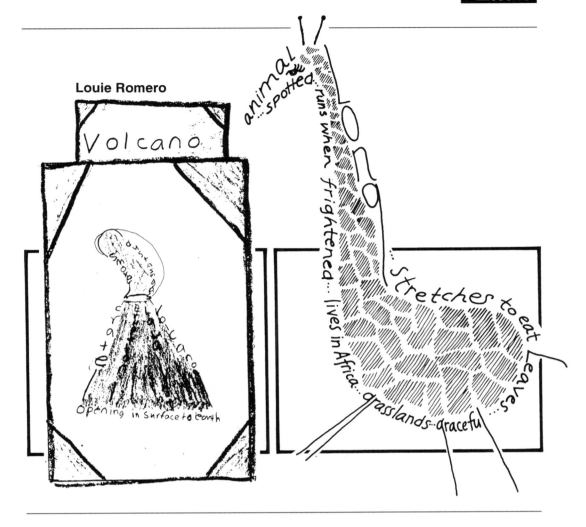

Louie Romero

Volcano

Assessment

These early lessons enable you to assess students' observation skills. The following criteria may help you assess written and oral descriptions:

- Has the student described observable properties?
- Is there evidence that the student has used more than one sense to observe?
- Has the student communicated observations completely and clearly?

In addition, these questions may help you evaluate student performance:

- How well does the student follow a written set of directions to perform a task?
- How well does the student work cooperatively in a group?
- Is the student able to complete a task with minimal help from the teacher?

Student Instructions for Taking a Sample of the Unknowns

1. Put on your goggles. Set up your tray with the test mat (use the side with the color names on it) and put the wax paper on top of the test mat.

2. Take out the red unknown sample jar and red measuring spoon. Open the jar.

3. Take one measure of the red unknown. Use your toothpick to level the amount on the spoon.

4. Put the sample on the red circle on your test mat.

5. Wipe off the spoon with a paper towel. Put the closed jar and spoon back in your science pail.

6. Observe the red unknown. Repeat Steps 2 through 6 with the four other unknowns.

Instructions: 1. Cut out both test mats along heavy outline.   2. Glue them back-to-back.   3. Laminate or cover them with plastic wrap.

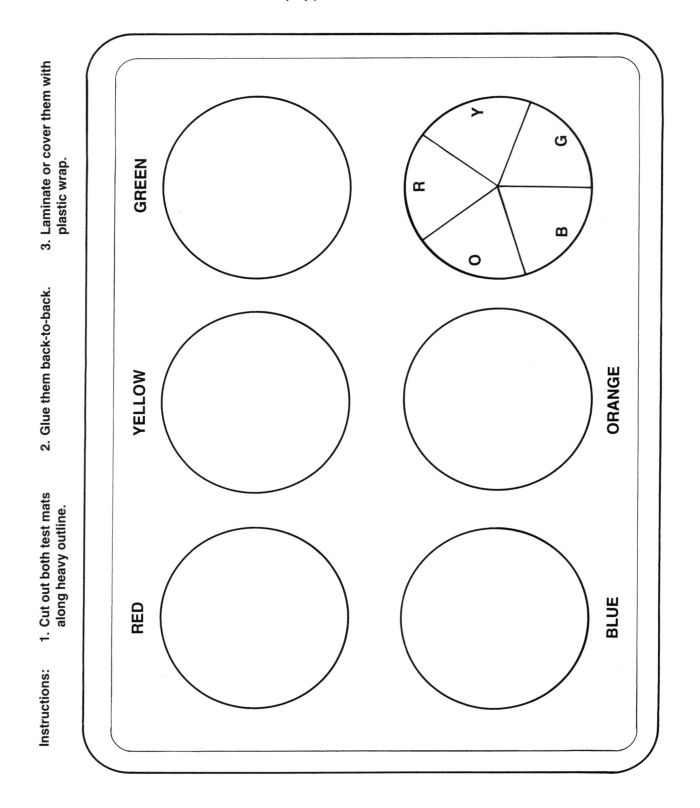

GREEN

YELLOW

RED

ORANGE

BLUE

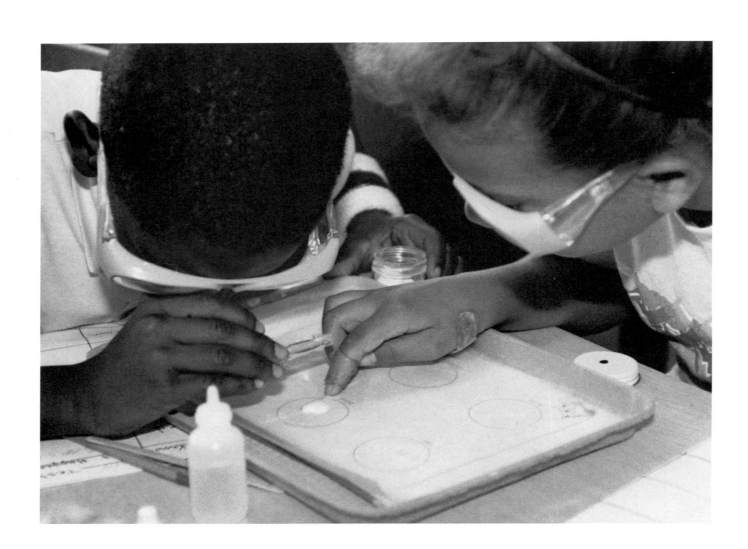

Testing Unknown Solids with Water

Overview and Objectives

In the last lesson, students explored some of the physical properties of the unknown solids. Over the next three lessons, they will continue this exploration by investigating the property of solubility. They do this by creating mixtures with each of the five unknowns and increasing amounts of water. This experience shows students that combining one substance with another can reveal new properties of each.

■ Students predict what they think might happen when they add a few drops of water to each unknown.

■ Students investigate the effects of mixing each of the unknowns with water.

■ Students record, organize, and discuss their results.

Background

In this lesson, students discover what happens when they add a few drops of water to a sample of each unknown. In effect, students are creating mixtures. The term **mixture** is to describe a combination of two or more different substances, each of which retains its own properties and can readily be separated from the other. (For example, salt water is a mixture of salt and water that can be separated by evaporation of the water.) Figure 4-1 shows the test results for adding six drops of water to each unknown.

Figure 4-1

Water test table

Water Test Table

Color Code	Unknown	Results
Red	Sugar	Dissolves; liquid is clear
Yellow	Alum	Dissolves; liquid is clear
Green	Talc	Does not mix easily; water beads up; skin forms on water surface
Blue	Baking soda	Dissolves; liquid may be slightly cloudy
Orange	Cornstarch	Dissolves; liquid may be slightly cloudy

When they have finished testing, students may observe that an unknown has "melted" or seems to have disappeared. After they have explained these observations in their own words, you may want to introduce the term **dissolve.** When a material is dissolved (for example, sugar dissolved in water), it is spread evenly throughout another substance and is not visible.

Materials

For each student

   1   science notebook
   1   **Record Sheet 4-A: Test Results Table**
   1   pencil

For every two students

   1   science pail
   1   dropper bottle of water, 7 ml (¼ oz)
   1   "water" label for dropper bottle
   1   tray
   1   test mat
   2   paper towels
   1   sheet of wax paper, 16 x 22 cm (6 x 8½")
   5   toothpicks

For the class

   1   plastic funnel
   1   sheet of newsprint and a large marker
       Source of water
       Cleanup supplies

Preparation

1. Affix a "water" label to each dropper bottle.

2. Using the funnel, fill the 15 water bottles. Place them in the materials center.

3. Add the water bottle strip to the "Check Your Science Pail" poster.

4. Have student helpers cut and trim the wax paper for the trays.

5. Make sure the following items are posted:

   ■ "How We Are Finding Out about the Unknowns" list (from Lesson 3)

   ■ "Class Properties Table" (from Lesson 3)

6. On a sheet of newsprint make a "Class Prediction Table" (see Figure 4-2). Leave the "Results" column blank.

7. Make a copy of **Record Sheet 4-A: Test Results Table** for each student.

Procedure

1. Focus students' attention on the "How We Are Finding Out about the Unknowns" list. Ask them for some other ways—besides observing the unknowns—to find out more about them. If students do not suggest "adding something to the unknowns," ask what they did in Lesson 1 to find out more about the mystery chemical in the plastic bag. Explain that, similarly, today they will add water to the five unknowns. Have students record in their notebooks what they predict might happen when they add a few drops of water to each unknown.

Figure 4-2

Sample table

CLASS PREDICTION TABLE

| RED | Sugar   dissapered in water
Salt   Sugar |
| YELLOW | Soap powder
Drimilk |
GREEN	ajax body bowpen
BLUE	starch flour
ORANGE	baking soda powder

2. Pass out **Record Sheet 4-A** to each student. Point out that it lists the unknowns in the same order as the "Class Properties Table." Review Figure 4-3 on how to fold and label the table (Figure 4-1 on pg. 18 of the Student Activity Book). Explain that students will record the results from the water test in this lesson. They will use the chart again to record the results from a second test in the next lesson.

   Note: Explain that in a later lesson students will use all of their test results and notes to solve several mysteries, so it is very important for them to keep their writing small and clear and to make their entries in the correct boxes. Let students know they do not need to write in complete sentences but that they should record detailed observations.

3. Now review the **Student Instructions for Doing the Water Test** at the end of this lesson (pg. 20 of the Student Activity Book). Explain that the second step (wiping off the measuring spoon) will help avoid contamination. Also explore with students why it is important to use the measuring spoons accurately. Talk about how to use the toothpick to level off the amount of each unknown.

4. Have teams pick up their materials and begin the water test. Circulate around the room to make sure students are testing correctly and working well together.

5. Have students clean up. If necessary, suggest that they refer to Lesson 3 for cleanup reminders.

6. Ask students to write in their notebooks what they have found out by adding water to the unknowns.

Figure 4-3

Folding and labeling your test table

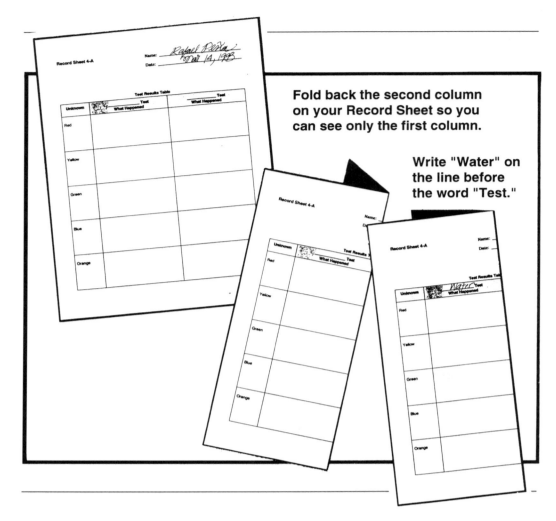

Fold back the second column on your Record Sheet so you can see only the first column.

Write "Water" on the line before the word "Test."

Note: In Step 7, the class will share test results. They will do this repeatedly during the remaining lessons. By asking questions such as the following, you can help students sharpen their observation skills:

■ What happened when you added water to the red unknown?

■ What did the water test tell you about each unknown?

■ What did you notice about the unknown when you added the water? What did it remind you of?

■ How are your results similar to those of your classmates? How are they different?

7. Have the class share observations. If students report that an unknown "disappeared," this is a good time to define the term **dissolve.** Assure students that there are no right or wrong observations; each observation is true for the student who makes it. Ask them to describe similarities and differences among the results.

8. Ask students to compare the predictions in their notebooks with the results on their test tables. Were they surprised by the results? Ask students to share their thoughts with the class.

9. Ask students to discuss what they now know about the unknowns that they did not know before. Then have students put their test tables in their notebooks to use in the next lesson.

Final Activities

1. Display the "Class Prediction Table." Explain that students can use it to record predictions about the unknowns' identities and to give reasons for their predictions. As students gather more evidence, they can add new ideas. (To encourage students who may be hesitant about appearing wrong, tell them not to write their names next to their predictions.)

2. Ask students to record in their own notebooks what they think the unknowns could be and why they think so, as well as any other thoughts they have as they continue testing.

Extensions

1. Have students sit in a circle on the floor and ask each student to place one shoe in the middle of the circle. Begin to sort the pile according to one property of the shoes. Ask students which property you are sorting by. Then put the shoes back in one pile and challenge students to choose another property to sort by.

2. Give every two students a clear cup of water. Have them predict what they think might happen when they put a drop of food coloring in each cup. Add the food coloring and have students record their observations as they watch the interaction. Encourage them to use words that describe what is happening as well as the properties of the materials. Remind them not to record opinions or comparative phrases, such as "It looks like ink."

3. Have students make a set of science safety posters to remind them of important safety rules (see Figure 4-4).

Figure 4-4

Sample safety poster

Rochelle Cooper

Assessment   Review the record sheets that students used to record results of the water tests. Are students using language that clearly describes the interactions of the substances? (For example, "The unknown disappeared into the water.") If students' written observations are not clear, ask them to tell you what they have observed. Then you will know whether they simply find it hard to transfer their experiences to paper or whether they are not observing closely.

Student Instructions for Doing the Water Test

1. Set up your tray, test mat, and wax paper. With the red measuring spoon, take one sample of the red unknown. Use your toothpick to level the amount on the spoon. Put the unknown in the red test circle.

2. Wipe off the measuring spoon with a paper towel before you put it back in the spoon bag. Put the unknown jar back in your science pail.

3. Put six drops of water on the red unknown sample. What happens? Use your hand lens for a closer look.

4. Use a toothpick to mix the water with the unknown. (Be careful not to rip the wax paper.) Now what happens? Record your observations on the water test table.

5. Repeat Steps 1 through 4 for the four other unknowns.

Record Sheet 4-A

Name: _____

Date: _____

Test Results Table

Unknown	_____Test What Happened	_____Test What Happened
Red		
Yellow		
Green		
Blue		
Orange		

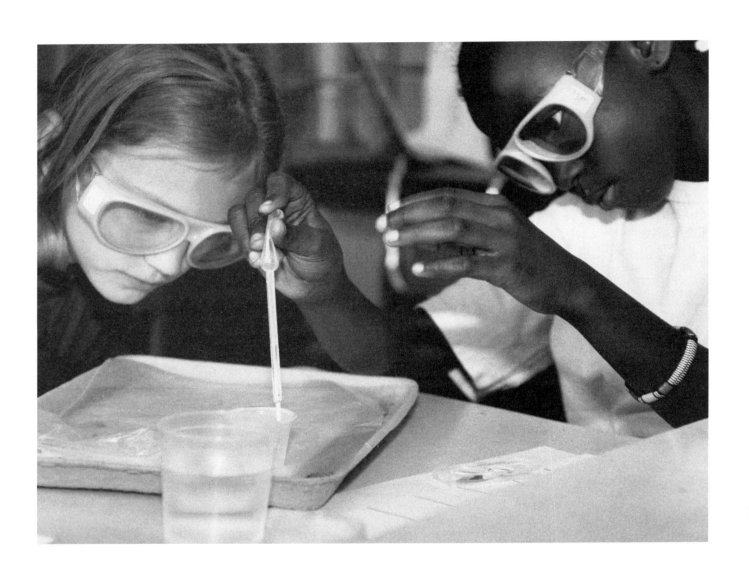

LESSON 5

Exploring Water Mixtures

Overview and Objectives

Students mix the five unknown solids with a greater amount of water than they used in Lesson 4. Through observations and discussion, students discover another property of the solids: some dissolve in water to form solutions, while others form suspensions. Through the process of filtering mixtures, students discover that solids in a suspension can be separated while those in a solution cannot. (In the next lesson, they will examine how to separate the solids in a solution: through the process of evaporation.)

- Students predict and then investigate what will happen when a greater amount of water is mixed with the unknown solids.

- Students observe and record the results of their investigations.

- Students filter the mixtures to explore further the physical properties of the solids.

- Students discuss their observations after filtering the liquid mixtures.

Background

In this lesson, students begin to investigate two types of mixtures: solutions and suspensions. A **solution** is a homogeneous mixture. Its components are uniformly mixed; therefore, you cannot see the parts that compose it. Solutions tend to be clear. For example, when you mix sugar with water, you create a solution. The sugar, which is the dissolved substance, is called the solute, and the water is called the solvent.

In this lesson, students gradually add small amounts of each unknown to a fixed amount of water. As students add more and more of some of the unknowns, the mixtures will reach a point at which no more of these unknowns will dissolve. When the water can hold no more of an unknown, it is said to be a saturated solution. Students will observe that some of each unknown is dissolved and some remains visible in the water. The remaining, or undissolved unknown is said to be suspended in the water.

A **suspension** is a heterogeneous mixture. Since its components are not uniformly mixed throughout, they can easily be seen. Suspensions are cloudy. For example, when you mix cornstarch with water, you create a suspension.

You can separate some suspensions simply by allowing them to remain undisturbed. The force of gravity causes the heavier material in the suspension to settle slowly to the bottom of the container. For example, think about salad

STC / *Chemical Tests*

59

dressing, a suspension containing oil and vinegar. Since the two ingredients separate naturally, you must shake the dressing before you use it to mix the parts more evenly.

In addition, you can separate some suspensions by filtration. **Filtration** is the process of passing a liquid through a porous substance in order to separate any solids from that liquid. Think about straining spaghetti and hot water. The substance that is suspended (spaghetti) is trapped by the filter, and the substance in which it is suspended (water) passes through the filter. When students filter their mixtures, some unknowns will be left in the filters.

Solutions, on the other hand, cannot be separated by filtration; the solution simply passes through the filter. Many solutions, however, can be separated by evaporation. Students will discover that when the water in their solutions evaporates, something is left.

Figure 5-1 illustrates the results you generally can expect from the water mixture tests in this lesson.

Figure 5-1

Water Mixture Test Table

Color Code	Unknown	Results
Red	Sugar	Dissolves; liquid is clear; forms a solution
Yellow	Alum	Dissolves; liquid is clear but with some unknown on bottom of cup; forms both a solution and a suspension
Green	Talc	Does not mix easily; does not dissolve; water appears white; forms a suspension
Blue	Baking soda	Dissolves; liquid is clear but with some unknown on bottom of cup; forms both a solution and a suspension
Orange	Cornstarch	Does not dissolve; liquid appears white; forms a suspension

This is the first lesson in which students use a control. As they create mixtures and witness the changes that occur, students learn that a **control** (in this case, water) is a reference point, an unchanged basis for comparison. The use of a control helps students make clearer, more accurate observations.

Management Tip: This lesson does not lend itself to being taught in two parts. Your students will need to create the water mixtures and filter them within the same time period.

Materials

For each student

1 science notebook
1 **Record Sheet 4-A: Test Results Table** (from Lesson 4)
1 pencil

For every two students

1 science pail
1 tray
1 test mat
5 plastic graduated cups, 37 ml (1¼ oz)
10 colored dots (2 red, 2 orange, 2 blue, 2 green, 2 yellow), .5 cm (¼")
1 plastic dropper
1 large plastic cup, 207 ml (7 oz)
2 paper towels
1 sheet of wax paper, 16 x 22 cm (6 x 8½")
5 toothpicks
5 evaporation dishes (tops and/or bottoms of petri dishes)
5 No. 1 coffee filters

For the class

1 red dot, .5 cm (¼")
1 sheet of newsprint or poster board and a large marker
1 gallon of water at room temperature
1 plastic graduated cup, 37 ml (1¼ oz)
Cleanup supplies

Preparation

1. Use the gallon container of room-temperature water to fill each of the 15 large plastic cups.

2. Place a red dot next to the 10-ml line on one of the small plastic graduated cups. Fill the cup with 10 ml of water (see Figure 5-2).

Figure 5-2

Place a dot next to the 10-ml line

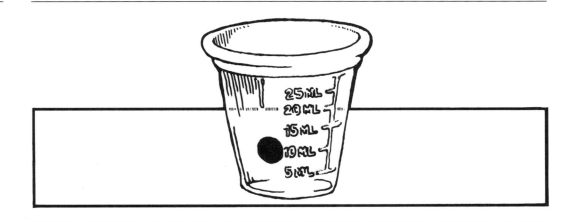

3. Have student helpers cut and trim 15 sheets of wax paper for the trays. Add these to the materials center along with the 15 large cups of water, 75 small cups, and 15 plastic droppers. (You will distribute the evaporation dishes and filters later.)

4. Cut apart the sheets of dots so that each team can pick up two dots of each color. Add these to the materials center.

5. Set up a classroom space where students can leave their trays of evaporation dishes and wet filters at the end of the lesson. Remember that placing dishes near natural light, air flow, and warmth will speed the rate of evaporation. Use your results from the evaporation test you set up at the end of Lesson 2 to help you choose a space.

6. Put the filters and evaporation dishes in an accessible area so you can distribute them to the teams after you have discussed filtering. (Separate the tops and bottoms of the petri dishes to use both for evaporation.)

7. On newsprint or poster board, prepare the headings for a class "Water Mixture Results Table" (see Figure 5-3).

Figure 5-3

Procedure

1. Hold up your small plastic cup of water. Ask the students what they think might happen if they mixed each unknown with more water than the six drops used in Lesson 4. Have them record their predictions in their notebooks. Let students know they will continue to explore the interaction of water and each unknown to see whether they can discover any new properties of the unknowns.

2. Have students take out **Record Sheet 4-A** from their notebooks. Ask them to label the second column "Water Mixture." Explain that this is where they will record today's test results.

3. Have teams pick up their materials. Then focus students' attention on the small plastic cup of water. Point out that it is marked with five different measurement units and explain that the class will be using the milliliter (ml) unit during this test.

    Note: These cups show measurement in drams (dr), cubic centimeters (cc), milliliters (ml), ounces (oz), and teaspoons (tsp)/tablespoons (tbs). See **Extension 4** on pg. 64 for ideas on exploring these different units of measurement.

4. Review the **Student Instructions for Doing the Water Mixtures Test** on pg. 65 of this guide (pg. 25 in the Student Activity Book). Answer any procedure-related questions. Then have students begin testing.

5. Circulate while students are working and check on the following details:

    ■   Make sure students use **Record Sheet 4-A** to record brief answers to the questions in Step 8 of the **Student Instructions.**

    ■   Make sure they compare the mixtures with the control cup of plain water to help them observe any changes.

    ■   Encourage students to place the black paper behind the cups and to use the hand lenses to enhance their observations.

    ■   Encourage students to hold the mixtures up to natural light to see them better.

6. When students have finished observing the mixtures, focus attention on the "Class Water Mixture Results Table." Looking at the questions in Step 8 of the **Student Instructions,** ask students briefly to share observations. Record them on the newsprint. Save the class table for Lesson 6. Ask the class how comparing the mixtures with the plain water in the large cups was helpful.

7. Now have students compare the predictions in their notebooks with their results. Ask them to discuss their findings with the class.

8. Ask students how they think the unknowns could be separated from the water. (Past student suggestions have included scooping out the unknowns, using tweezers, pouring off the water.)

    ■   If no one suggests filtering, ask how they separate a mixture of spaghetti and water at home. This should spark the idea of a filter or strainer.

    ■   Hold up a filter and ask if students have ever seen one used. Then explain that the class will use it to try to separate the mixtures. (Ask students to put **Record Sheet 4-A** in their notebooks.)

9. Distribute five filters and five small dishes to each team. (Remember, do not yet refer to them as "evaporation dishes.")

10. Go over the **Student Instructions for Filtering the Water Mixtures** on pg. 67 of this guide (pg. 27 in the Student Activity Book). Have students get to work.

Final Activities

1. When students have finished, show them where to put their trays of dishes and filters. Remind them to walk slowly when they carry the trays.

2. Discuss with students what they have observed during filtration. Then explain that they will let the trays sit for two or three days. They will observe the filters and dishes daily. Tell them to design their own filtration results tables in their notebooks and to record observations there.

3. Clean up. Collect the used plastic cups and have students wash them.

4. Now have students write in their notebooks what they have learned by creating mixtures of the unknowns and water. Ask questions such as the following:

- What did you learn by creating mixtures of the unknowns and water?

- In what ways were the five water mixtures you created similar?

- In what ways were the mixtures different?

- What were some properties of each mixture you observed?

Management Tip: Do not begin Lesson 6 until evaporation is complete. (This should not take more than two or three days.) Use the results from the evaporation test you set up at the end of Lesson 2 to gauge how long evaporation will take. Plan observation times accordingly. If evaporation was rapid, you may want to have students observe more than once a day.

Extensions

1. Set up a filtration learning center so students can test some of their ideas about filtration. Have them filter a variety of mixtures, such as tea, coffee, and salad dressing.

2. In a quart jar (a mayonnaise jar works well), create "dirty" water with soil, sticks, and leaves. Challenge your students to design a method to clean the dirty water and to describe this method in writing. If possible, let them actually test their designs.

3. Read a book such as *The Magic School Bus at the Waterworks*, by Joanna Cole, to the class (see **Bibliography** for reference). This book illustrates the water cycle and water purification process and discusses the role of alum (the yellow unknown).

4. The gradations on the small plastic cups used in this lesson reflect five different measurements. Ask the class to investigate how the measurement systems differ, how they have evolved, and how each is used today.

Student Instructions for Doing the Water Mixtures Test

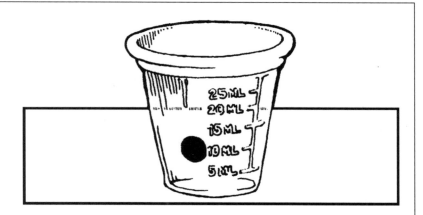

1. Find the 10-ml line on one small cup and put the red dot next to the line. Then color-code the other four cups.

2. Put the red cup on the red circle on the tray. Your teacher will show you how to fill the plastic dropper by squeezing the bulb, placing it in the large cup of water, and releasing the bulb.

3. Fill your plastic dropper. Hold it right over the small cup and squeeze the bulb. Repeat this until the water comes up to the 10-ml line on the cup. To make sure the water level is right, have your partner view the cup at eye level as you add the water.

4. Using your red measuring spoon, add one measure of the red unknown to the cup. Stir the mixture with a toothpick for 30 seconds.

5. Add two more measures of the red unknown **one at a time.** After each measure, stir again for 30 seconds. (You now have added a total of three measures.)

6. Observe the mixture you have just made.

7. Let the cup sit on the tray. Repeat Steps 2 through 6 for the four other unknowns. Switch jobs with your partner so that you each get a chance to use the dropper.

8. Record on **Record Sheet 4-A** short answers to the following questions:

- Where is the unknown in the water? Can you still see it?

- What does the water mixture look like? Compare it with the plain water left in your large cup. Has it changed? If so, how?

- How does stirring affect the way the mixture looks?

Student Instructions for Filtering the Water Mixtures

1. With your partner, use a pencil to write your names near the top of each filter. Under your names, label one filter red, one orange, one green, one blue, and one yellow. Then color-code the dishes by placing one dot on the side of each dish.

2. Carefully move the small cups from the tray to your desk. Make sure to keep the cups away from the edge of the desk and away from your arm.

3. Put the red dish on the tray. Pick up the red filter and use both hands to hold it open directly over the dish.

4. Have your partner stir the water mixture in the red cup once with the toothpick and then slowly pour it into the filter. Observe what happens.

5. Wait until no more liquid is dripping out of the filter. Leave the dish undisturbed on the test mat. Put the filter near the tray on your desk.

6. Repeat Steps 3 through 5 for each mixture.

Discovering Crystals

Overview and Objectives

In this lesson, students observe dramatic changes that have occurred in the filtration dishes from Lesson 5: the liquid in the dishes has evaporated, and crystals have appeared in the dishes that contained solutions. Through the experiences in Lessons 5 and 6, students discover that, depending on an unknown's solubility in water, a solid can be separated from water by either evaporation or filtration. Students continue to learn that some changes take place over periods of time and that observing these changes can give them new information about the properties of the five unknowns.

- Students predict, observe, and discuss the filtration results of the mixtures they created in Lesson 5.

- Students record their observations on a test table of their own design.

- Students continue to explore the properties of two types of mixtures, solutions and suspensions.

- Students read and write about crystals.

Background

In Lesson 5, students filtered their water mixtures. They then let them sit for a few days. Depending on how soluble each unknown is in water, when evaporation is complete, you generally can expect the results shown in Figure 6-1. Results may vary because of classroom conditions and the degree to which students followed directions.

In this lesson, students observe the crystals that formed from the alum and sugar. A **crystal** is a solid substance with a regular geometric pattern (a repeating arrangement of its atoms). When a crystal grows, layer is added to layer, repeating the original geometric pattern. Crystal sizes vary according to the conditions under which they are formed. For example, the longer the formation period, the larger a crystal will be.

Many students find the study of crystals interesting. They often want to learn more about them. Let your students know that they will learn more about crystals in a reading selection at the end of the lesson.

When sharing results, students may ask why the yellow unknown now looks like a crystal and not a powder. Ask for their ideas. Some students have said it was a crystal because "something happened" when it mixed with the water. This is partially true. Because of evaporation, the unknown, which already is composed of tiny crystals, reappears as larger crystals. **Evaporation** is the process by which a liquid (in this case, water) changes into a gas (in this case, water vapor).

Figure 6-1

Filtration Test Results Table

Color Code	Unknown	Results
Red	Sugar	Filter paper empty; crystalline sticky mass in dish; separated by evaporation
Yellow	Alum	Filter paper has some residue; crystals in dish; separated by evaporation and filtration
Green	Talc	Filter paper full; dish empty; separated by filtration
Blue	Baking soda	Filter paper has some residue; white film and some crystals visible in dish; separated by evaporation and filtration
Orange	Cornstarch	Filter paper full; dish empty; separated by filtration

All five unknowns are actually tiny crystals; however, some are so small that they cannot be seen as crystals, even with the aid of a magnifying lens. Extension 2 on pg. 75 provides electron microscope photographs that will help students see what each unknown looks like under very high magnification.

Materials

For each student
  1   science notebook
  1   pencil
  1   **Record Sheet 6-A: Growing Sugar Crystals**
  1   **Record Sheet 6-B: Growing a Giant Crystal**

For every two students
  1   tray with five dishes (from Lesson 4)
  5   filters (from Lesson 4)
  1   science pail

For the class
  1   "Class Water Mixture Results Table" (from Lesson 5)
  1   sheet of newsprint
  5   colored markers (red, yellow, green, blue, orange)
  1   container of soapy water
  1   container of rinse water
      Cleanup supplies

Preparation

1. Using the newsprint, set up a table entitled "Class Filtration Test Results" (see Figure 6-2). Post the table.

2. Hang the "Class Water Mixture Results Table" from Lesson 5 next to the "Class Filtration Results Table."

Figure 6-2

Sample table

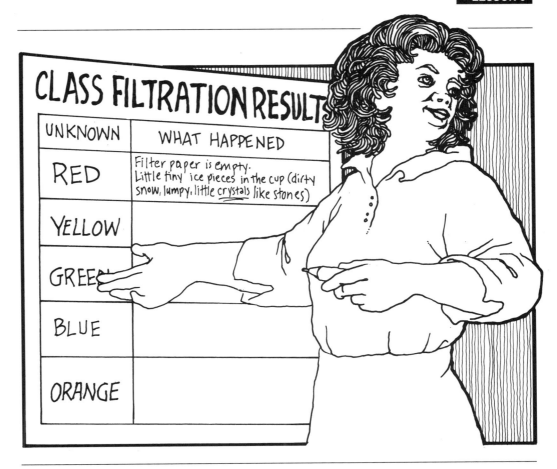

CLASS FILTRATION RESULTS

UNKNOWN	WHAT HAPPENED
RED	Filter paper is empty. Little tiny ice pieces in the cup (dirty snow, lumpy, little <u>crystals</u> like stones)
YELLOW	
GREEN	
BLUE	
ORANGE	

3. Make photocopies of **Record Sheet 6-A: Growing Sugar Crystals** and **Record Sheet 6-B: Growing a Giant Crystal** (pgs. 78–79) for each student.

Procedure

1. Have the students open their notebooks to the filtration results tables they designed in Lesson 5. Ask teams to share the observations they recorded.

2. Have students pick up their science pails and the trays with the dishes and five filters.

3. Review the following steps with the class. (These steps also appear on pgs. 29–30 in the Student Activity Books.) Pair each team with another team for the fourth step. Then have students observe and record their filtration results.

   ■ Observe the dish and open the filter paper for each unknown.

   ■ Record your results for each unknown in the filtration results table in your notebook.

   ■ Compare what is left in the filter paper and dishes with the untested samples in the unknown jars. Have the unknowns changed? In what ways?

   ■ Discuss your results with another team. Talk about the answers to the following questions:

     ▪ What happened to the mixtures after filtration?

     ▪ Are your team's results similar to those of the other team?

     ▪ If your results are different, why do you think that happened?

     ▪ Which results surprise you? Why?

4. Now focus the students' attention on the "Class Filtration Results Table." Ask teams to share their results. Record the most common results for each category on the chart. (See Figure 6-2 on pg. 73 for typical class results for the red unknown.) To help the class discuss possible reasons for varied results, ask the following questions for each unknown, beginning with the red one:

   ■ What did you observe about the filter paper used with the red unknown water mixture? About the dish?

   ■ What do you think happened to the water? Where is the unknown?

   ■ How do you think the water and unknown were separated? (Students probably will mention evaporation. Take a moment to discuss this.)

5. Summarize which unknowns were separated by filtration, which by evaporation, and which by both by doing the following:

   ■ Focus attention on the "Class Water Mixtures Results Table" from Lesson 5. Point out that the orange and green unknowns (and, in some instances, some of the yellow unknown and the blue unknown) were still visible in the water. Now look at the "Class Filtration Results Table" and ask how the orange and green mixtures were separated. Write "separated by filtration" in the results column for these unknowns.

   ■ Next, refer to the "Class Water Mixtures Results Table" for those mixtures in which none of the unknown remained visible (red) or some—but not all—of the unknown remained visible (yellow and blue). Referring to the "Class Filtration Results Table," ask students how these mixtures were separated. Write "separated by evaporation" or "partly by evaporation" in the corresponding results column.

Final Activities

1. Take a few minutes to ask questions such as the following:

   ■ What do we now know about the unknowns that we did not know before?

   ■ In what ways were the five mixtures you created similar? In what ways were they different?

   In the discussion, help students arrive at the understanding that those mixtures in which the unknowns were still visible in the water were separated by filtration. Those mixtures in which the unknowns were not visible in the water were separated by evaporation. Those mixtures in which some unknown remained visible were separated by both filtration and evaporation.

2. Clean up. Collect the crystals that grew from the yellow and red unknowns to use in the **Extensions.** Put the other evaporation dishes in the container of soapy water for students to wash later. Ask students to throw away the filters.

3. Read "A World of Crystals" as a whole class, in small groups, or in pairs.

4. Have students write in their notebooks three things they have learned about crystals. Also ask them to record any questions they may still have. Discuss with students how they might find out the answers.

5. Hand out copies of **Record Sheets 6-A** and **6-B** to each student. Ask students to choose one of the crystal-related activities to do at home. Explain that an adult must be present when they do these activities. Have students observe their experiments for several days and record observations in their notebooks. Then hold a class discussion about the results.

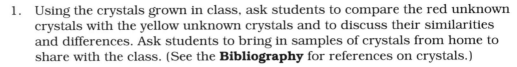

Extensions

1. Using the crystals grown in class, ask students to compare the red unknown crystals with the yellow unknown crystals and to discuss their similarities and differences. Ask students to bring in samples of crystals from home to share with the class. (See the **Bibliography** for references on crystals.)

   Note: Save the crystals of red and yellow unknowns for use in Extension 4 in Lesson 12.

2. Copy the blackline master of the electron microscope photographs of the five unknowns on pg. 80 for each student. Explain to the class that an electron microscope is an instrument that is capable of very high magnification. Then have students compare how each unknown looks through their hand lenses (which magnify 3x and 6x) with the photographs that illustrate the unknowns magnified either 100x or 1000x. Challenge your students to calculate how many more times the crystals were magnified by the electron microscope.

3. Challenge students to separate a mixture of salt water and one of sand and water.

4. Read books such as *Two Bad Ants*, by Chris Van Allsburg, to the class. This is a story of two ants' attraction to sugar crystals. (See the **Bibliography** for references.)

5. Collect geometric solids such as cubes, prisms, cylinders, pyramids, and cones. Have students sort the solids by their attributes. Ask students in what ways they are similar to or different from the crystals they have observed.

6. Have the class do research on how silicon chips help run computers, VCRs, and other electronic equipment. Or ask students to write letters to a computer company and ask questions about this topic. If possible, invite a computer technologist to talk to the class.

Assessment

Review individual record sheets for the water mixture and filtration tests.

■ Are students recording their observations and results in a descriptive, informative, and organized manner?

■ Did students describe different properties of the mixtures they observed?

■ Did they accurately and completely describe in what ways the mixtures changed as a result of filtration and evaporation?

Reading Selection

A World of Crystals

Did you know that crystals are all around you? Outside, they make up rocks, minerals, snow, and the sand that crunches between your toes at the beach. Inside, your home is filled with crystals. In your freezer are ice crystals. In your Mom's jewelry box, there may be gems or metals that are crystals. You sprinkle crystals of sugar on your cereal and mix crystals with water to make lemonade.

A sparkling diamond. A snowflake. A speck of sugar or a grain of salt. These are all crystals. They seem like very different things. But in some ways, they are the same. How? All crystals are solid. They are made up of pieces. And those pieces are laid out in a pattern. That pattern is repeated over and over again.

All crystals of the same material have a similar shape. Take quartz, for example. You can look at a piece of quartz the size of a

Quartz crystal

pea or one the size of a pumpkin. The quartz can be from Africa—or from right here in the United States. No matter what, the shape of the quartz crystal is always similar.

Cubes on Your French Fries?

Some crystals are so small that you must look through a microscope to see them. Then you'd know their shapes. Did you know you were eating tiny cubes on your french fries? Or that if you could catch a snowflake and see it under a microscope, it would have six points?

Thanks to a famous scientist named Roger Bacon, crystals help us take a closer look at the world. A long time ago, Bacon discovered that by shaping and polishing quartz, he could make a lens. This lens made small objects look bigger. And that was the start of modern eyeglasses, microscopes, telescopes, and even contact lenses.

Today, it's hard to imagine living without these things. But back then, people thought that lenses were some kind of black magic and that Roger Bacon was a wizard. They actually put him in jail for his important discovery.

Snowflakes

Many Uses

Crystals are valuable in other ways. For ages, kings and queens have worn crowns made of precious metals such as gold and silver. Pirates lined their chests with jewels such as red rubies, green emeralds, and diamonds.

In fact, diamonds are not only beautiful to look at, but they are also very useful. Because they are so hard, they can be used to saw, drill, grind, and polish. So, we use diamond saws to cut microchips for computers. We use diamond dust to polish glass and other hard things.

Today, we have many more uses for crystals, especially a part of quartz called "silicon." Silicon helps run and control computers, calculators, and microwave ovens. Also, quartz helps make some watches work.

Crystals are a mystery, in a way. They are not alive, and yet they grow. Want to see how? Ask your teacher about how you can grow some crystals at home.

Silicon helps run computers, microwaves, and watches.

Record Sheet 6-A

Name: _____

Date: _____

Growing Sugar Crystals

Safety Note: Only do this activity when an adult is present.

Materials

Water, 236.5 ml (1c)
Granulated sugar, 236.5 ml (1 c)
Measuring cup
Small saucepan
Wooden spoon
A glass or jelly jar
Cotton string, about 20 cm (8″)
Pencil
Stove (source of heat)
Small weight (a nut, small washer,
paper clip, or button)

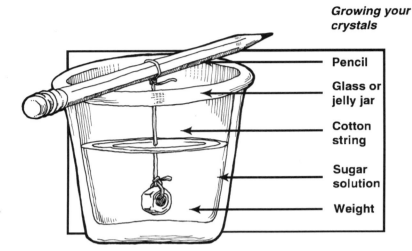

Growing your crystals

- Pencil
- Glass or jelly jar
- Cotton string
- Sugar solution
- Weight

Find Out for Yourself

1. Make sure all of your equipment is clean.

2. Fill the pan with one cup of water.

3. When the water boils, turn off the heat.

4. Add a spoonful of sugar right away. Stir. Keep adding sugar and stirring. When some sugar sits on the bottom of the pan even after you stir it, stop adding sugar.

5. Let the sugar water cool.

6. Pour the sugar water into the glass.

7. Wet the string. Tie the small weight on the end of the string.

8. Tie the other end of the string around the middle of the pencil. (When you hang the pencil over the glass, the small weight should hang just above the bottom of the glass.)

9. Put the pencil across the top of the glass and hang the string in the sugar water.

10. Put the glass in a safe place where it will not be disturbed.

11. Let the glass stand for three to five days. Check it every day and remove crystals that form on the surface of the water so that the water can continue to evaporate. Watch what grows on the string!

12. With your adult partner, discuss what happens.

13. Then try this again using Epsom salt or salt!

STC / Chemical Tests

Record Sheet 6-B

Name: _____

Date: _____

Growing a Giant Crystal

Safety Note: Do this activity only when an adult is present.

Materials

> Cold water, 5 tbs
> Salt, 5 tsps
> Mixing cup
> Saucer
> Tweezers

Find Out for Yourself

1. Mix the 5 teaspoons of salt with the 5 tablespoons of cold water in the mixing cup.

2. Pour the mixture into the saucer.

3. Let the saucer sit undisturbed. Observe what happens over the next few days.

4. Make a new salt solution by repeating Steps 1 and 2.

5. Take out the biggest crystal with the tweezers and put it in the new salt solution.

6. Repeat this again and again until you have a giant crystal.

Note: To get crystals of different shapes, try using washing soda, Epsom salt, or sugar. Each will dissolve in boiling water. To make colored crystals, add food coloring to the water before you start!

Blackline Master
Electron Microscope Pictures

Pictures of the five unknowns are magnified either 100 or 1000 times to allow for easy observation of the individual particles.

Photos reproduced with the permission of Dr. Andrew MacInnes and Mr. Christopher Landry, Department of Chemistry, Harvard University

Red unknown: magnified 100x.

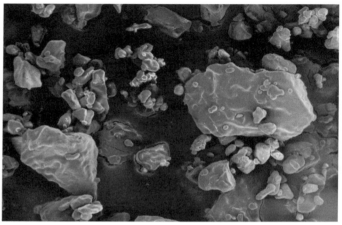

Yellow unknown: magnified 100x. Each particle contains many small crystals.

Green unknown: magnified 1000x.

Blue unknown: magnified 100x.

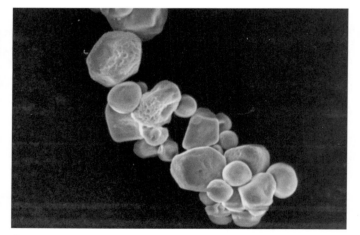

Orange unknown: magnified 1000x. Each particle contains many crystals.

Testing Unknown Solids with Vinegar

Overview and Objectives

Over the past four lessons, students have explored the physical properties of the unknown solids. In each of the next three lessons, they will use three different chemical liquids—vinegar, iodine, and red cabbage juice—to investigate the unknowns' chemical properties. Beginning with vinegar, students will discover how adding each of these different liquids to each unknown solid causes different chemical reactions that produce changes in the solid's form, color, or texture. In later lessons, students will use this data to determine how such properties might be used to identify the unknowns. In addition, students will start to understand the value of using a control when observing the results of chemical tests.

- Students predict the results of testing the five unknowns with vinegar.

- Students test the unknowns and record and discuss their observations of the different reactions.

- Students discuss how the use of a "compare circle," or control, helps them interpret test results.

- Students write their thoughts on what they have learned by testing the unknowns with vinegar.

Background

In this lesson, students begin to examine the unknowns' chemical properties. A **chemical property** of a substance is its ability to transform into new materials. We can observe chemical properties when substances react with one another. For example, when iron is exposed to moist air, a new substance (rust) may form. So, the ability to rust is a chemical property of iron. The ability to burn is a chemical property of wood and cotton.

In this lesson, students will observe a chemical property of the blue unknown (baking soda). When vinegar is added to the baking soda, the two substances react chemically to form bubbles of carbon dioxide. In general, you can expect the vinegar test to produce the results in Figure 7-1.

In Lesson 5, students compared how plain water looked before and after they had added the unknown. Although they may not have been aware of it at the time, students were using the water as a **control**—an unchanged basis of comparison—to make these observations. In this lesson, students are formally introduced to the concept of a control when they use the unlabeled circle on their test mats to hold a small, untested sample of each unknown.

Figure 7-1

Vinegar Test Table

Color Code	Unknown	Results
Red	Sugar	Dissolves
Yellow	Alum	Dissolves
Green	Talc	Does not dissolve; vinegar beads up; skin forms on vinegar; liquid is cloudy
Blue	Baking soda	Dissolves; fizzes and bubbles
Orange	Cornstarch	Does not dissolve; liquid is cloudy

Materials

For each student

  1 science notebook
  1 pencil
  1 **Record Sheet 7-A: Test Results Table**

For every two students

  1 science pail
  1 dropper bottle of vinegar, 7 ml (¼ oz)
  1 tray
  1 test mat
  1 sheet of wax paper, 16 x 22 cm (6 x 8½")
  5 toothpicks
  1 paper towel

For the class

  1 "How We Are Finding Out about the Unknowns" list (from Lesson 3)
  1 plastic funnel
  1 stock bottle of vinegar, 250 ml (½ pt)
     Cleanup supplies

Preparation

1. Put vinegar labels on the 15 dropper bottles and fill them with vinegar. Place the bottles in the materials center.

2. Add the vinegar bottle strip to the "Check Your Science Pail" poster.

3. Have student helpers cut and trim 15 sheets of wax paper for the materials center.

4. Copy **Record Sheet 7-A: Test Results Table** for each student.

5. Display the "How We Are Finding Out about the Unknowns" list.

Procedure

1. Focus students' attention on the "How We Are Finding Out about the Unknowns" list. Ask students what to add. (They should suggest the water test results and water mixture test results.)

2. Ask students what they know about vinegar. Also ask them to predict what might happen if they added a few drops of vinegar to each unknown. Have students record ideas in their notebooks. If there is time, have the students share their ideas.

3. Hand out **Record Sheet 7-A.** Ask students to fold back the second column and write "Vinegar" in front of the word "Test" at the top of the first column. Tell them to record all observations on this table.

4. Introduce the use of a "compare circle" by asking students how they will be able to tell whether the unknowns change after vinegar is added. If they do not suggest comparing the vinegar mixture with a plain sample of the unknown, remind them how they used the plain water in the water mixture test. Then ask: "How would having a sample of each plain unknown on your tray help you describe the results of the vinegar test more accurately?"

   Explain that students will use the sixth circle on their test mats as a "compare circle"—a place to put a small, untested sample of each unknown that they can compare with the tested sample. Point out that a scientist would call the samples in the compare circles the "controls."

5. Go over the **Student Instructions for Doing the Vinegar Test** on pg. 88 of this guide (pg. 37 in the Student Activity Book).

6. Ask if there are any questions about the testing procedure. In Lessons 9 and 10, students will follow the same instructions using different liquids. Be sure they understand how to use the compare circle.

7. Have students pick up their materials and begin the vinegar test.

8. Circulate around the room. Check for the appropriate use of the compare circle. Talk with the teams about how they are using it.

9. Have students clean up their materials and workspace. Remind them to look at the "Check Your Science Pail" poster.

Final Activities

1. Have students share results of their vinegar tests. Review their observations for each unknown. Ask questions such as the following:

   ■ How did vinegar react with each unknown? In what ways were the reactions similar? Different?

   ■ How did the untested samples in the compare circle help you describe your observations?

   ■ How were the vinegar test results similar to those of the water test? How were they different?

   ■ How did your predictions compare with your results?

2. Have your students record in their notebooks what they have learned by adding vinegar to the unknowns (see Figure 7-2). If they would like, have them add to the "Class Prediction Table."

Figure 7-2

Sample
notebook entry

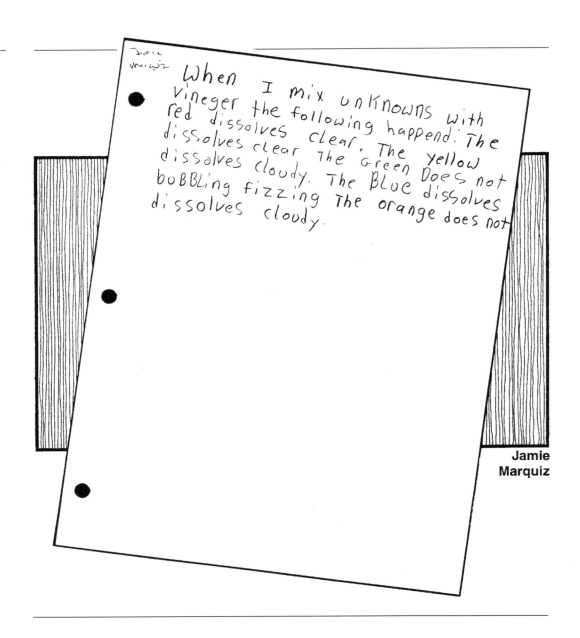

When I mix unknowns with vineger the following happend. The red dissolves clear. The yellow dissolves clear The Green Does not dissolves cloudy. The Blue dissolves buBBling fizzing The orange does not dissolves cloudy.

**Jamie
Marquiz**

Extensions

1. For a related art activity, make mosaics out of powdered objects. You need glue, salt, sand, nondairy creamer, food coloring, and dark construction paper.

   ■ Color the salt, sand, and creamer with different food colorings.

   ■ Have students use lines of white glue to design a pattern or scene on the paper.

   ■ Have students sprinkle the creamer and colored salt and sand on the glue lines wherever they choose.

   When students are finished, discuss how the properties of the materials created the mosaic.

2. Students may have seen vinegar in their kitchens at home. Refer to the **Bibliography** for books such as *Science Experiments You Can Eat*, by Vicky Cobb, and have students choose some "kitchen chemistry" experiments to do at school or at home.

3. Have each student create a winter scene. First, have them cut out shapes from colored construction paper and add details with crayons. Next, tell them to glue the shapes onto a dark piece of paper. Mix 250 ml of Epsom salt or alum (the yellow unknown) with 125 ml of water. Tell students to use a brush to paint the solution over the picture. Let the picture dry thoroughly and wait for crystals to appear. Ask students to describe how the solution changed the picture.

Assessment

The **Assessment** section on pg. 103 in Lesson 9 contains guidelines you may find useful over the next few lessons. You may want to refer to these guidelines at this time.

Note: In Lesson 9, students will test the unknowns with red cabbage juice. Because the juice requires advance preparation on your part, please read the **Background** and **Preparation** sections on pgs. 99 and 100 at this time.

Student Instructions for Doing the Vinegar Test

1. Set up your tray, test mat, and wax paper.

2. Using the red measuri[n] spoon, take a very sma[ll] amount of the red unknown. Put it in the "R" section of the compare circle. This is your control.

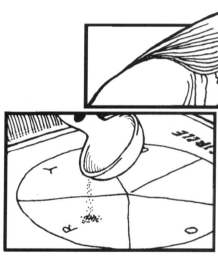

3. Using the red measuring spoon, take one full sample of the red unknown. Use a toothpick to get a level spoonful and put it in the red circle.

   ■ Wipe the spoon with a paper towel. Put it back in the spoon bag.

   ■ Put the red unknown jar back in your science pail.

4. Put six drops of vinegar on the sample of red unknown in the red circle.

- Wait a few seconds. What happens?

- Look at the sample in the compare circle. Compare it with the unknown in the red circle.

- Did the unknown change after vinegar was added? If so, how? Record your observations on your "Test Results Table."

5. Use a toothpick to mix the vinegar and the red unknown. What happens now? Record any new observations.

6. Repeat Steps 2 through 5 for the four other unknowns.

Record Sheet 7-A

Name: _____

Date: _____

Test Results Table

Unknown	_____Test What Happened	_____Test What Happened
Red		
Yellow		
Green		
Blue		
Orange		

| LESSON 8 | **Testing Unknown Solids with Iodine** |

Overview and Objectives

Building on their experiences in Lesson 7, students test the unknowns with a second chemical liquid—iodine. They continue to observe that mixing one substance with another may produce changes in form, color, or texture. By following the same test procedure used in Lesson 7, students reinforce skills in testing, observing, using a control to make comparisons, and recording and interpreting results.

- Students predict the results of testing the five unknowns with iodine.

- Students test the unknowns and record and discuss their observations.

- Students continue to learn safety guidelines for working with chemicals.

Background

Figure 8-1 shows the results you generally can expect from this test.

Figure 8-1

Iodine Test Table

Color Code	Unknown	Results
Red	Sugar	Dissolves; liquid is color of iodine
Yellow	Alum	Dissolves; liquid is color of iodine
Green	Talc	Does not dissolve; iodine beads up
Blue	Baking soda	Dissolves; liquid is color of iodine
Orange	Cornstarch	Mixture turns purple-black

Since iodine turns purple-black in the presence of starch, mixing the orange unknown and iodine produces a very quick, dramatic effect. Keep in mind that wax paper contains a little starch, and the iodine will react very slightly with the wax paper after a few minutes of contact. If, after several minutes, students observe reactions with the other unknowns, avoid telling them that the iodine is reacting with the starch in the wax paper. Simply explain that chemicals are everywhere—including in the wax paper and in the iodine—and that when they combine, chemicals sometimes cause reactions. Also point out the contrast between these slight reactions and the dramatic reaction of the orange unknown.

Safety Note: This unit uses highly diluted iodine (0.1%), which is not harmful if ingested in small quantities. Iodine, however, **is** considered toxic when ingested in large quantities. If a student ingests a large quantity, call your local poison control center immediately.

Additional Notes

- Iodine can stain paper and clothes. To remove stains, soak clothes in a mixture of vitamin C and water.

- Iodine is sensitive to both air and light. It can evaporate and lose its strength. For this reason, the iodine solution in this kit comes in an amber stock bottle. Keep the iodine in this container until you need to fill the dropper bottles for the iodine test.

- The iodine solution in the dropper bottles may lose its indicator properties after a few weeks. If this happens, replace the contents of the dropper bottles with iodine solution from the stock bottle. Reminders to check the strength of the iodine solution appear at appropriate places throughout the unit.

Materials

For each student
1   science notebook
1   **Record Sheet 7-A: Test Results Table** (from Lesson 7)
1   pencil

For every two students
1   science pail
1   dropper bottle of 0.1% iodine, 7 ml (¼ oz)
1   tray
1   test mat
5   toothpicks
1   sheet of wax paper, 16 x 22 cm (6 x 8½")
1   paper towel

For the class
1   plastic funnel
1   stock bottle of 0.1% iodine solution, 250 ml (½ pt)
    "How We Are Finding Out about the Unknowns" list
    Cleanup supplies

Preparation

1. Put iodine labels on 15 dropper bottles and fill them with the iodine solution. Add the dropper bottles of iodine to the materials center.

2. Add the iodine bottle strip to the "Check Your Science Pail" poster.

3. Have student helpers cut and trim the 15 sheets of wax paper and add them to the materials center.

Figure 8-2

Add the vinegar test to the list

HOW WE ARE FINDING OUT
ABOUT THE UNKNOWNS

1. Observing them
2. Water test
3. Water Mixture test
4. Vinegar Test

Procedure

1. With the class, add the vinegar test to the "How We Are Finding Out about the Unknowns" list.

2. Ask students what they know about iodine and what they predict might happen if they add it to the unknowns. Have students record their ideas in their notebooks. Then explain that today students will add a new liquid— iodine—to each of the five unknowns.

3. Review relevant safety rules from the "Safety Rules" poster. Be sure to point out that it is important to clean up all spills immediately.

4. Have students take out **Record Sheet 7-A: Test Results Table** and unfold the second column. Tell them to write "Iodine" in front of the word "Test."

5. Let students know that the iodine test uses the same procedure as the vinegar test. If they need to review the test procedure, suggest that they turn back to pg. 37 in the Student Activity Book.

6. Have students pick up their materials and perform the test.

7. Ask students to clean up their workspaces and return their materials.

Final Activities

1. Have the class share results. For each unknown, ask what changes students observed after adding the iodine. Then ask if any of the unknowns reacted differently from the others. (Most students will name the orange unknown.) Talk about the fact that changes can occur when they mix one substance with another. You may want to cite other examples students may have seen, such as iron rusting when it was exposed to moist air, tea changing color when lemon juice was added to it, or the blue unknown fizzing when vinegar was added to it in Lesson 7.

2. Now, ask students to compare the predictions in their notebooks with the results on their test tables. Which results did they find surprising? Ask students to share their thoughts with the class.

3. Have students record in their notebooks answers to the following questions:

   ■ What have you learned about the unknowns by doing the iodine test?

   ■ Look at all your test results so far. Have any unknowns reacted similarly when you used different tests?

Extensions

1. Place a drop of iodine on a microscope slide and spread it out with a toothpick. Allow it to dry. Have students observe the slide under a microscope. Breathe on the slide. What happens to the crystals?

2. Have students write their own mystery stories. Ask what they think makes a good mystery. What do they like about the mystery books they have read? Remind students to include those elements in their own stories.

3. Write a class mystery story. Have students sit in a circle. On newsprint, write a line or two to start the mystery story. Then roll a small rubber ball to a student. He or she will add the next line or two. That student then rolls the ball to another until the class finishes the story.

Testing Unknown Solids with Red Cabbage Juice

Overview and Objectives

In this lesson, students use a third chemical liquid, red cabbage juice, to explore further the chemical properties unique to each unknown. Following the completion of this test, students analyze all the data they have collected over the past seven lessons and summarize what they now know about the physical and chemical properties of each of the five unknown solids. This is a good time for the class to add to the "What We Think about Chemicals" list from Lesson 1.

- Students predict the results of testing the five unknowns with red cabbage juice.

- Students test the unknowns and discuss and record their observations.

- Through individual writings and class discussion, students reflect on what they have learned about the unknowns from the tests they have conducted so far.

Background

When students test the unknowns with red cabbage juice, you generally can expect the results shown in Figure 9-1.

Figure 9-1

Red Cabbage Juice Test Table

Color Code	Unknown	Results
Red	Sugar	Dissolves; liquid is color of juice
Yellow	Alum	Dissolves; liquid is deeper, brighter purple
Green	Talc	Does not dissolve; liquid is color of juice (slightly whiter)
Blue	Baking soda	Dissolves; liquid turns green
Orange	Cornstarch	Does not dissolve; liquid is color of juice (slightly whiter)

Red cabbage juice is a **natural indicator;** that is, a substance that, through its color, reveals the presence of certain chemicals. Because this is a plant juice and contains no preservatives, its shelf life is short. BHT, a nontoxic food preservative, has therefore been added to the red cabbage juice in this kit.

Freshly made red cabbage juice has a blue-purple tint. However, after a while, it may assume more of a bluish tint. This slight color change will not interfere with its indicator properties.

Your students may not be as familiar with red cabbage juice as they are with water, iodine, and vinegar. For your information, a property of red cabbage juice is that it is an acid-base indicator. Students will discover this property in Lesson 15, when they test a variety of household chemicals with the juice.

Materials

For each student
- 1 science notebook
- 1 **Record Sheet 9-A: Test Results Table**
- 1 pencil

For every two students
- 1 science pail
- 1 dropper bottle of red cabbage juice, 7 ml (¼ oz)
- 1 tray
- 1 test mat
- 5 toothpicks
- 1 sheet of wax paper 16 x 22 cm (6½ x 8″)

For the class
- "How We Are Finding Out about the Unknowns" list (from Lesson 3)
- "What We Think about Chemicals" list (from Lesson 1)
- "What We Would Like to Know about Chemicals" list (from Lesson 1)
- 1 marker
- 1 stock bottle of red cabbage juice, 250 ml (½ pt)
- Cleanup supplies

Preparation

1. Check the color of the red cabbage juice.

    ■ If the juice is green, add vinegar a drop at a time until it is blue-purple.

    ■ If the juice is pink, add baking soda a small pinch at a time until it is blue-purple.

2. Test the juice's indicator properties by adding a few drops to samples of the yellow and blue unknowns. The mixtures should turn purple and green, respectively. If the color change is faint, make a fresh batch of juice. (See **Appendix D** for directions.)

3. Put red cabbage juice labels on 15 dropper bottles and fill them with red cabbage juice. Place them in the materials center.

4. Add the red cabbage juice bottle strip to the "Check Your Science Pail" poster.

    Note: Although this is the last item you will add to the poster, continue to leave it up. Students will refer to it in later lessons.

5. Have student helpers trim 15 sheets of wax paper for the trays. Add them to the materials center.

6. Copy **Record Sheet 9-A: Test Results Table** for each student.

7. Be sure the "What We Think about Chemicals" and "What We Would Like to Know about Chemicals" lists are accessible.

Procedure

1. Introduce the subject of testing red cabbage juice by asking questions such as the following:

   ■ What do you know about red cabbage or red cabbage juice?

   ■ What do you predict might happen if you add red cabbage juice to the unknowns?

   Have students record ideas in their notebooks and discuss them with the class. Then explain that students will test the unknowns with another liquid: red cabbage juice.

2. Hand out **Record Sheet 9-A.** Tell students to fold back the second column and write "Red Cabbage Juice" before the word "Test" at the top of the first column.

3. Explain that students will use the same procedure for this test as they did for the vinegar and iodine tests. There will be only one difference: after students add the drops of cabbage juice to each unknown, they should count slowly to 15 before observing the unknown and recording their observations. If students need to review the test instructions, remind them to turn back to Lesson 7, pg. 37, in the Student Activity Book.

4. Ask the teams to pick up their materials and begin the red cabbage juice test.

5. Have students clean up their materials and workspaces.

6. Have teams share their results with the class by asking the following questions:

   ■ How did the red cabbage juice react with each unknown? What were the similarities and differences among the reactions?

   ■ How did the compare circle help improve your observations?

   ■ Did any unknowns react differently or unusually from the others? (Blue and yellow generally produce the most dramatic results.)

   ■ How did your predictions compare with the results?

Final Activities

1. Add the iodine and red cabbage juice tests to the "How We Are Finding Out about the Unknowns" list. Use the list to help the class review the tests that have been done so far.

2. Have students record in their notebooks thoughts about questions such as the following:

   ■ What do you now know about each unknown that you did not know before testing?

   ■ In what ways are the unknowns similar? In what ways are they different?

   ■ What do you know about chemicals now that you did not know before? What new questions do you have?

3. Focus students' attention on the "What We Think about Chemicals" list. Using a new color of marker, add students' latest thoughts to the list. If students have any new questions, add them to the "What We Would Like to Know about Chemicals" list.

Figure 9-2

Sharing results

Extensions

1. Read books to the class such as *The Legend of the Indian Paintbrush,* by Tomie DePaola. In this Native American story, plant juices are used for painting. Have students research plants used by Native Americans to dye cloth.

2. Use the following procedure to make colored paints from plant juice dyes. Recipes for specific colors follow the procedure.

   - Put the plant parts into a saucepan and cover with cool water.

   - Boil for the amount of time indicated below.

   - Strain the juice into a glass jar. Let it cool.

   - Mix each dye with a paint medium or with modeling clay base. You may also add it to cotton string to make weavings.

   Recipes

   - Red coloring: Use beets (cut up) or cranberries. Cook 30 minutes.

   - Blue-purple coloring: Use red cabbage. Cook 15 minutes.

   - Yellow coloring: Use onion skins. Cook 30 minutes.

   - Green coloring: Use spinach leaves. Cook 30 minutes.

   - Blue coloring: Use cornflower petals (cut in small pieces) or blueberries. Cook 15 minutes.

   - Brown coloring: Dissolve one teaspoon of instant coffee in two tablespoons of hot water.

3. Ask how many students have eaten red cabbage. Find recipes that use it. Or bring in a red cabbage for everyone to taste.

4. Use red cabbage juice and other plant dyes (see preceding directions) to tie-dye T-shirts and other clothes with the class.

Assessment

This is a good time to check students' progress in the following areas.

Class and Team Discussions

■ Can students orally describe what they are doing and observing when testing the unknowns?

■ When discussing results, do students refer to the samples in the compare circles? For example, "When I added the juice to the yellow unknown, it got wet and purple. It wasn't white and dry like it is in the compare circle."

■ Are students using new vocabulary to describe their observations and results?

■ Are students becoming aware that different chemicals have different properties?

■ Are students articulating an understanding that changes may occur when different chemicals are mixed together or separated?

Record Sheets

■ Can students express in writing what they have observed?

■ Are students' recorded observations becoming more descriptive? Is there a progression from, for example, "It looked wet" to "It got wet and darker in color, and after I mixed it, it was all liquid."

■ Do the record sheets indicate that students are beginning to refer to the compare circle when observing results?

Notebook Entries

■ Can students describe their experiences testing the unknowns?

■ Are students identifying dramatic or different test results? For example, "When I tested the blue unknown with vinegar, it was the only one that bubbled a lot."

■ Are students beginning to identify similarities and differences between the properties of the unknowns? For example, "They all mixed with the water, but only the red and yellow dissolved much."

■ Compare the notebook entry from Step 2 of the **Final Activities** in this lesson with the first notebook entry from Lesson 1.

  ■ Are students beginning to see that testing the unknowns helps reveal new properties of each?

  ■ Are students demonstrating growth in what they think they know about chemicals as well as developing awareness that everyday materials are made of chemicals?

Note: Allow time for those students who have difficulty writing to discuss their work with you.

Testing Skills

- Are students following testing instructions more accurately?

- Are students becoming more comfortable with handling the materials?

- Are students becoming more responsible about cleaning up?

Record Sheet 9-A

Name: _____

Date: _____

Test Results Table

Unknown	_____Test What Happened	_____Test What Happened
Red		
Yellow		
Green		
Blue		
Orange		

Testing Unknown Solids with Heat

Overview and Objectives

In Lesson 9, students analyzed their data from five tests. These tests involved adding water and other chemical liquids to the five unknown chemical solids. In this lesson, students continue to broaden their concept of chemistry and their laboratory skills by investigating the effects of heat on the solids. They discover that heat produces dramatic changes in the form, color, odor, and texture of some of the unknowns. They also continue to develop good safety practices.

■ Students predict the results of testing the five unknowns by heating them.

■ Students test the unknowns and record and discuss their observations.

■ Students discuss and use new safety guidelines when heating materials.

Background

Many students are fascinated by the heat test—because of both the heating process itself and its dramatic results. While the test requires close supervision by the teacher, keep in mind that it is specifically designed to ensure your class's safety.

As one safeguard, students will rotate through the heat station in manageable teams of five. Because you will repeat the test several times, you may want to spread it out over two days. You may also want to enlist the help of an adult volunteer.

While each team is performing the test, engage the others in activities such as making safety posters about the use of heat, reading mystery books, or illustrating unit activities they have enjoyed. If students remain distracted by the activity, that is fine; it will help them understand what to do when it is their turn at the station.

In general, you can expect the heat test results in Figure 10-1.

Management Tip: The setup and procedure in this lesson outlines one way for a class of 30 to do the heat test. You may want to consider other methods (for example, setting up a single heat station for each unknown) for your particular class.

Figure 10-1

Heat Test Table

Color Code	Unknown	Results
Red	Sugar	Liquifies; turns brown and syrupy; smokes; turns black; smells like candy
Yellow	Alum	Snaps and crackles; bubbles turn hard and remain white
Green	Talc	Does not react much; turns gray
Blue	Baking soda	Remains white and powdery; may give off smoke
Orange	Cornstarch	Smokes; turns brown and black; smells like burned toast

Materials

For each student

1   science notebook
1   **Record Sheet 9-A: Test Results Table** (from Lesson 9)
1   pair of goggles

For every five students

5   aluminum bake cups (with paper liners removed)
5   wooden clothespins
1   foil-lined tray

For the class

1   aluminum bake cup (candle holder)
1   pie tin, 22.9 cm (9″), with water
1   votive candle
1   "Safety Rules" list
5   toothpicks
5   measuring spoons
5   jars of unknowns
    Cleanup supplies

For the teacher

1   box of safety matches
1   jar for used matches
1   pair of goggles
1   marker
1   science pail
1   roll of tape
1   roll of aluminum foil

Figure 10-2

Heat station

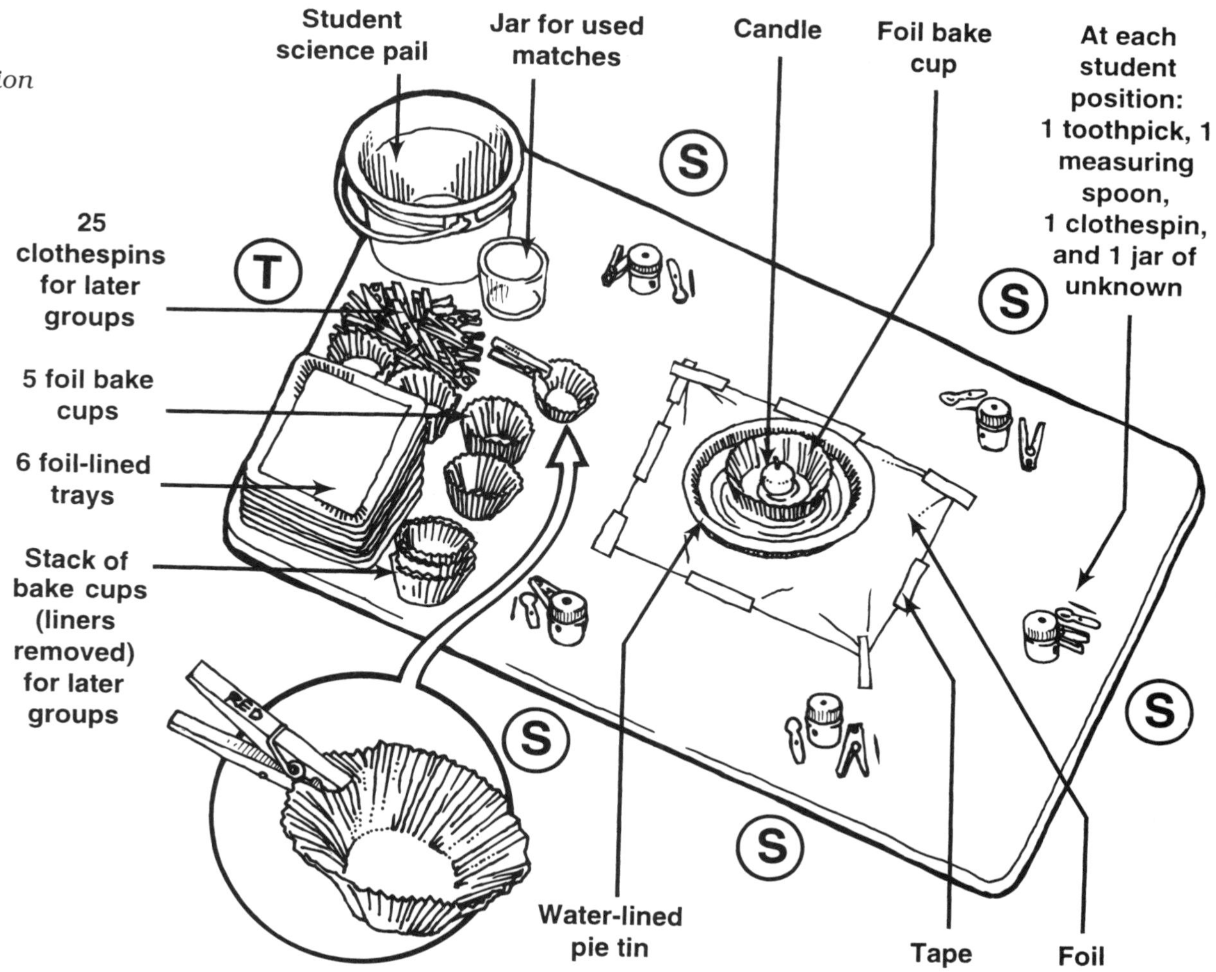

Student science pail

Jar for used matches

Candle

Foil bake cup

At each student position: 1 toothpick, 1 measuring spoon, 1 clothespin, and 1 jar of unknown

25 clothespins for later groups

5 foil bake cups

6 foil-lined trays

Stack of bake cups (liners removed) for later groups

Water-lined pie tin

Tape

Foil

Preparation

1. Line six trays with sheets of aluminum foil.

2. Set up a heat station away from the flow of student traffic and in the area least likely to distract students when they are not directly involved in the heat testing (see Figure 10-2).

   ■ Place one science pail of unknown samples and measuring spoons at the station. Also put the foil-lined trays, bake cups, clothespins, and toothpicks at the station. They will be distributed to each group of five students.

   ■ Attach a clothespin to one of the bake cups (see inset, Figure 10-2).

3. Designate a follow-up station area where each team of five can discuss test results.

4. Display the "Safety Rules" list in an accessible location.

5. Carefully read the procedures. Divide the class into teams of five and list each team's members on the chalkboard. This will speed things along when you are ready to switch groups at the heat station. Select a helper from each team. Make sure the team helper is a good reader.

6. Have ready some activities students can do while awaiting their turns at the heat station.

   Safety Notes

   ■ Be sure all students, and you, are wearing goggles before you light the candle.

   ■ Be sure to keep your goggles on until all teams have completed the heat test.

   ■ Tie back long hair (that is, hair below the earlobe).

   ■ Roll up long sleeves.

   ■ Make sure the workspace around the candle is free of paper.

Procedure

1. Ask students what has happened when they have watched materials being heated on a kitchen stove or over a campfire. What do they predict might happen if they heat the unknowns? Ask students to record ideas in their notebooks. Then explain that they will do one more test on the unknowns: they will add heat.

2. Have students take out **Record Sheet 9-A.** Have them unfold the second column and write "Heat" on the top in front of the word "Test." Discuss the safety issues outlined in the **Safety Notes** above. Add new concerns to the "Safety Rules" list. Have students get their goggles from their science pails.

3. Review the heating setup and the **Student Instructions for Doing the Heat Test** on pg. 112 of this guide (pg. 49 in the Student Activity Book). Point out why students will do this test one team at a time.

4. In addition, you will need to

   ■ Remind students to bring their notebooks, a pencil, and goggles to the heat station.

   ■ Remind team helpers to bring their Student Activity Books.

   ■ Explain that each student will be responsible for heating one unknown as well as for observing the heating of all five.

   ■ Make sure students do not lean on the heat stations when recording their observations.

   ■ When each team is ready to test, light the candle. Put the used match in the jar.

   ■ Blow out the candle at the end of each team's test.

   ■ After each team finishes the test, put a new clothespin and bake cup at every space for the next team.

5. Have only one team at a time at the heat station. Ask the students who are not at the heat station to work on the other activities you have planned.

6. As each team finishes testing, have its members take their notebooks and sit together at the follow-up station. Have the team helper carry the team's tray. Explain that the team helper will lead the group follow-up discussion by asking questions about each unknown. The questions, found in the Student Activity Book on pg. 48, are reprinted for you here:

   ■ What did you observe about the unknown when it was heated?

   ■ What do you observe about the unknown after it has cooled?

■ In what ways did the unknown change as a result of heating?

Have the team helper throw away the aluminum cups and return the tray and clothespins to the materials center. Instruct students to put their record sheets in their notebooks, return to their seats, and work on the other activities you have assigned.

7. Ask team helpers to help clean up the heat station.

Final Activities

1. Review the heat test results with the class. Focus on the range of results. Talk about the questions they discussed at the follow-up station. Ask how their predictions compared with their results.

2. Ask, "What did you discover from heating the unknowns that you didn't learn from adding liquid to them?" Have students record ideas in their notebooks.

3. Tell the students that this was their last test with the unknowns. In the next lesson, they will discuss ways to use all their test results to help solve the mystery.

Extensions

1. Chemicals can produce different colors when heated to their ignition point. Ask students to do some research on how fireworks work.

2. Suggest that students do research concerning which gases produce the different colors of fluorescent lights.

3. Read books such as *Strega Nona's Magic Lessons*, by Tomie DePaola (see **Bibliography** for reference). Using examples from the story, point out that the interaction of chemicals can produce change. For example, Strega Nona heats her potions to make them work. Light a candle and let the students list all the changes they can observe (for example, odor, smoke, blackening of the wick). Explain that when chemical changes occur, substances change to new substances with different properties.

Assessments

This is a good time to have the class fill out the Student Self-Assessment in Appendix A (see pg. 180). Also ask students to fill out the self-assessment after they complete the unit. By comparing earlier and later responses, students get a good sense of their individual growth.

Student Instructions for Doing the Heat Test

1. Go to your heat station space. Put your notebook and test table on your chair.

2. Write the color of the unknown on the handle of your clothespin. Your teacher will show you how to attach the clothespin to your bake cup.

3. When it is your turn, put one measure of the unknown at your heat station space in the bake cup your teacher gives you. You will heat this unknown. You will watch while your teammates heat the others.

4. When heating an unknown, hold the cup one inch above the flame. Make sure the sample in the bake cup is directly over the flame.

5. When you observe the unknown as it is being heated, do not lean directly over the unknown.

6. Without leaning on the table, record your test results. Repeat these steps for each unknown.

7. When you are finished, carefully place your bake cup on your team's tray.

Reviewing the Evidence

Overview and Objectives

Your students have carried out two physical tests (water drop and water mixtures) and four chemical tests (vinegar, iodine, red cabbage juice, and heat) on the unknown solids. In this lesson, they review and analyze all the data they have collected as a result of performing these tests to determine the distinctive properties of each unknown. This process provides students with the information and skills they need to solve the mystery in Lesson 12.

- Students create test summary tables that enable them to review the data they have collected on the unknowns.

- Students summarize and analyze their results in order to discriminate differences between the unknowns.

- Using all their test results, students predict the identities of the unknowns and provide reasons that support their predictions.

Background

Throughout this unit, your students have been gathering evidence to solve the mystery of the five unknowns. In this lesson, you will ask students to make individual judgments about which test results reveal important properties of particular unknowns.

The goal is to have each student highlight what is meaningful to him or her; therefore, there are no "correct" answers. Whether or not a student highlights the reaction of the blue unknown to vinegar, for example, is not significant. What is significant is the validity of the student's reason for finding a result important. Listen carefully as students describe what they have chosen to highlight (see the **Assessment** section on pg. 118 for helpful assessment criteria).

Materials

For each student
1 science notebook (containing completed copies of **Record Sheets 4-A, 7-A,** and **9-A**)
1 piece of white construction paper, 30 x 46 cm (12 x 18″)
1 pencil
1 scissors

For every two students
Glue

For the class

    1   model "Test Summary Table"

    4   assorted markers

        "Class Properties Table" (from Lesson 3)

        "How We Are Finding Out about the Unknowns" list (from Lesson 3)

        Cleanup supplies

Preparation

1. Add scissors, construction paper, and glue to the materials center.

2. Using blank copies of **Record Sheets 4-A, 7-A,** and **9-A,** make a model "Test Summary Table." The procedure for doing this is in Figure 11-2.

3. Post the "Class Properties Table" and "How We Are Finding Out about the Unknowns" list.

Procedure

1. With the class, add "Heat Test" to the "How We Are Finding Out about the Unknowns" list. Then review the six tests students have done so far. Ask them to take out **Record Sheets 4-A, 7-A,** and **9-A,** which contain the results of all six tests.

2. Ask students how they think they can use all their test results to identify the unknowns. Record their ideas on the chalkboard. If students need a motivator, you can tell them what detectives—and scientists—often do with the information they collect: they write all of it down in one place and then analyze it. Discuss how students can analyze their test results to identify the unknowns. See Figure 11-1 for other students' suggestions.

Figure 11-1

Sample list

3. Have the class look at the "Class Properties Table" and ask, "How could a table like this, showing all your test results, help you?" Then suggest that students each make a summary table containing all their test results.

4. Show the class the model "Test Summary Table" you have prepared. (It will be blank, but remind students that this is just a model.) Explain that each student will combine all of his or her test tables from **Record Sheets 4-A, 7-A,** and **9-A** to make a "Test Summary Table" like the model.

5. Review the steps for putting together the "Test Summary Table" (see Figure 11-2, which also appears as Figure 11-1 on pg. 52 in the Student Activity Book). Emphasize which record sheet to glue first, second, and third. Let students get to work. Remind them to clean up all scraps when they have finished.

6. When students have assembled their tables, explain that they will use a pencil or marker to highlight the test results they consider unique to each unknown—the results that help students distinguish one unknown from the others. They can highlight the result by circling it or by placing a star next to it.

7. As an example, have students look at their vinegar test results and highlight the one(s) that seemed especially significant. (Many students may highlight the results of the blue unknown, which bubbled.) Ask students to share their decisions and to give reasons for them. Emphasize that there is no "right" answer. Students should highlight any discoveries that seem unique to an unknown and that might help them identify it.

8. Have students begin working. As they finish, tell them to keep their test summary tables on their desks and to return their materials to the center.

9. Ask students, "Now that you have highlighted your important results, how can you use them to solve the mystery of the unknowns?" Suggest the idea of comparing the information students have collected with information about known chemicals in books or other reliable sources. Explain that students will have a chance to do this in the next lesson.

Final Activities

1. Ask a few volunteers to show their tables to the class and to explain which test results they have highlighted and why. (You may want to ask students who don't volunteer to share their tables with you privately later in the day.)

2. Have students record in their notebooks predictions of the identity of each unknown. Ask them to give at least two reasons for each prediction. You may want to write a sentence starter such as "I think the unknown is _____ because _____." (A likely response might be "I think the red unknown is sugar, because it dissolved in water and smelled like candy when we heated it.")

3. Have students fold their test summary tables and put them in their notebooks.

Extensions

The class can gain more experience with properties and change by making "sinking gel." Because this activity uses full strength ammonia, do it as a demonstration activity. You will need the following:

- ½ tsp of alum (Do not let students know that this is the yellow unknown!)
- 2 tsps of household ammonia
- 1 baby food jar half-filled with water
- 1 toothpick
- 1 large plastic cup, 207 ml (7 oz)
- 1 small plastic cup, 37 ml (1¼ oz)

Directions:

- Put the ammonia in the large cup. Put the alum in the small cup.

- Ask the students to observe the contents of both cups. Have the class list the properties of the ammonia and the alum.

- Add the alum to the water. Stir the mixture with a toothpick.

- Then add the ammonia and stir again.

- Have students observe the jar for about five minutes and record their observations (the solution will turn cloudy and, after standing, a white gel will start to settle to the bottom).

- Then have students list any new properties of the alum and ammonia they have discovered as a result of the test.

- Discuss the changes that occurred.

Assessment

From this lesson, you can assess each student's ability to interpret the importance of specific test results in identifying the unknowns. As you review their notebook entries, consider the following criteria:

- Are students highlighting test results that reveal properties unique to each unknown?

- Are they making comparisons between unknowns?

- Are they supporting their predictions with actual test data? (Remember, it is not important that students correctly predict the identity of the unknowns but that they provide plausible reasons to support their predictions.)

As students describe why they highlighted certain test results, think about the following questions:

- Are students correctly using new vocabulary and descriptive language?

- Are they looking for patterns in the results?

- Are students citing test changes that reveal new properties?

- Are students detecting differences between the unknowns?

Figure 11-2

Putting together the "Test Summary Table"

1. Trim the left margin of Record Sheet 4-A (water and water mixtures). Glue Record Sheet 4-A onto the far left side of the construction paper.

2. Take Record Sheet 7-A (vinegar and iodine). Cut off the column of color names.

3. Glue Record Sheet 7-A onto the construction paper, next to the water and mixtures table. Be careful to align the columns.

4. Take Record Sheet 9-A (cabbage juice and heat). Cut off the column of color names, and trim the right margin.

5. Glue Record Sheet 9-A next to the vinegar and iodine table. The right side will hang off the paper a little.

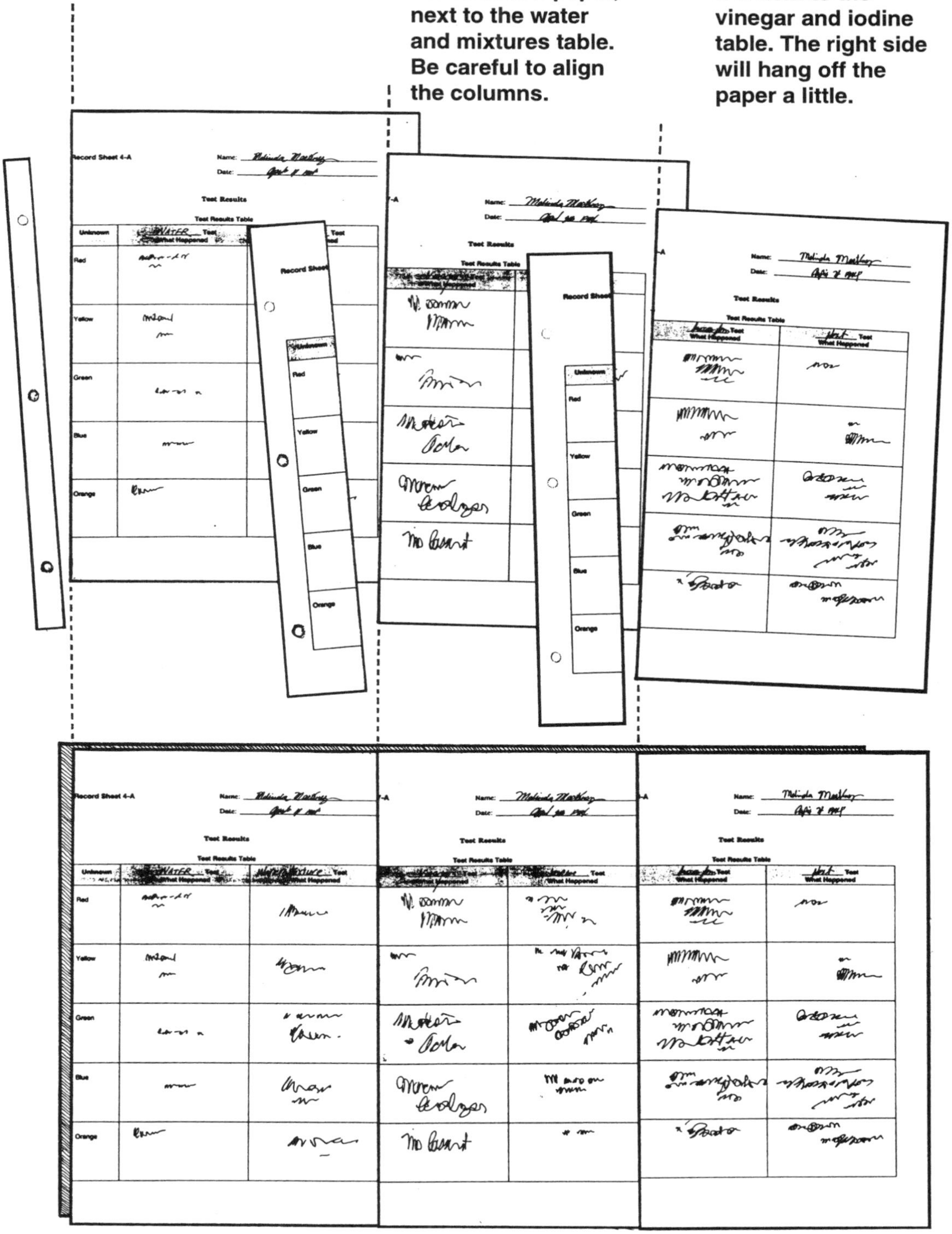

LESSON 12

Identifying the Unknown Solids

Overview and Objectives

By analyzing and discussing their test data in the last lesson, the students identified the unique physical and chemical properties of each unknown solid. In this lesson, students are challenged to identify each unknown by comparing their test data with verified test data from chemists.

- Students use the process of comparing their test data with a reliable source of information to identify the unknowns.

- Students read about how the five chemicals they have been testing are used in everyday life.

Background

How do scientists identify unknowns? What about an unidentified rock sample from the moon, for example? Scientists compare the moon rock's properties with those of known rocks on earth. By seeking correlations—and ideally, a match—scientists can determine the moon rock's composition.

In this lesson, students follow a similar process to discover the identities of their five chemical unknowns. They compare the data they have collected through observing and testing the unknowns (contained in their test summary tables from Lesson 11) with data from a reliable source of information (contained in **Record Sheet 12-A: Chemical Information Sheet**).

Materials

For each student
1. science notebook
1. pencil
1. **Record Sheet 12-A: Chemical Information Sheet**
1. **Reading Selection: "Chemicals Are All Around Us"**

For the class
  "Class Properties Table" (from Lesson 3)

Preparation

1. Copy **Record Sheet 12-A: Chemical Information Sheet** for each student.

2. Copy "Chemicals Are All around Us" (pgs. 125–26 of this guide) for each student.

3. Read "Chemicals Are All around Us." Decide if you want students to read independently, in small groups, or as a class.

   Note: You may want to do this lesson's **Final Activities** during language arts.

4. Hang the "Class Properties Table."

Procedure

1. Ask students to summarize their ideas about how their test results could help them identify the five chemical unknowns. Then explain that you will give each student a copy of **Record Sheet 12-A,** which describes the properties of some common chemicals. Ask how students might use this information to help them identify the unknowns.

2. Have students take out their test summary tables from Lesson 11. Distribute **Record Sheet 12-A.**

3. Now challenge the class to use their test summary tables, the "Class Properties Table," and the Chemical Information Sheet to find out the identities of the unknowns and solve the mystery. Point out where on the record sheet students will write in the color of each unknown. Encourage them to discuss their discoveries with one another.

4. When students have finished, go over **Record Sheet 12-A** as a class. Have students record in their notebooks their thoughts about the following questions:

   ■ Did any of the unknowns' identities surprise you? If so, why? If not, why not?

   ■ How did your predictions of the unknowns' identities compare with your discoveries?

   ■ Which test results were the best clues for helping you identify an unknown?

   ■ What do you now know about sugar that you did not know before? (Repeat this question for each of the other unknowns.)

Final Activities

1. Give each student a copy of "Chemicals Are All around Us" and have students read it. Afterwards, ask the class to talk about the common uses for the five unknowns.

   Note: You may want to share with the class that in nature, talc is a mineral with no scent. Why do we scent talc? Discuss this topic with your students.

2. Have students write and illustrate fictional accounts of what it would be like to live in a land without one or more of the five chemicals they have been studying (sugar, alum, baking soda, talc, and cornstarch).

3. Ask students to put their test summary tables and Chemical Information Sheets in their notebooks.

Figure 12-1

*Discovering the
identities of the
unknowns*

Extensions

1. Point out that there are many books in the library about chemicals and chemistry. Encourage students to try books such as *Everyday Chemicals*, by Terry Jennings, or *Chemistry for Every Kid*, by Janice Pratt Van Cleave (see the **Bibliography** for references).

2. Create an "Everyday Chemicals" bulletin board with your class. Have students draw or bring in pictures of common chemicals, label them, and describe their uses.

3. Discuss the safe use of household chemicals with the class. Emphasize proper storage and handling. Use the "Everyday Chemicals" bulletin board as a basis for your discussion.

4. Now that students have discovered the identities of the five unknowns, have them apply the chemical tests to the crystals grown in Lesson 6. Are the crystals really the red unknown (sugar) and the yellow unknown (alum)?

Assessment

Now that the students have discussed five unknown solids and read "Chemicals Are All around Us," the following questions may help you evaluate student growth:

- Are students becoming more aware that everyday materials are made of chemicals?

- Have students demonstrated an understanding that different chemicals have different properties and that some chemicals can be identified by their interactions with other chemicals?

- Are students developing an awareness of the importance of chemicals in our lives?

Reading Selection

Chemicals Are All around Us

Think about the five unknowns you tested. You may not realize it, but we use these chemicals every day!

Do you like pickles? Well, when pickles are made, one type of **alum** helps keep them so crisp that they crunch when you bite into them. Also, if you fall off your bike and cut your knee, alum may be in the medicine you put on it. It helps heal cuts and scrapes.

Alum has another important use: it helps purify water, making it clean and safe for you to drink. How does it work? When we add alum to a tank of dirty water, it sticks to the dirt and mud. This helps form globs that sink to the bottom. Then the clean water flows off the top.

Flat as a Tortilla?

Of course, **baking soda** is used for baking. Without it, biscuits, cakes, and muffins wouldn't rise—they'd look more like tortillas! Baking soda also cleans things. Try it on your toothbrush instead of toothpaste. And, if you're bitten by a mosquito or stung by a bee, have an adult put baking soda into your bath water. It will help take away the pain and itching, and your bite will heal faster.

Do you remember what happened when you added water to the green unknown—**talc?** The water drops beaded up and a film formed. The water wouldn't mix well with the talc. That's why parents like to use nice, soft, scented talcum powder under babies' diapers. Talc protects the babies' skin and helps keep them dry.

And, if parents run out of talcum powder, they can use **cornstarch** to powder the baby. (Remember, cornstarch didn't mix well with water, either. Your tests results were thick and sticky.) We also use cornstarch to make foods stick together when we cook. Without it, the gravy on your mashed potatoes might be watery, not thick.

You Can't "Beet" Sugar!

You already know that candy and other sweets contain **sugar.** But where does the

sugar come from? It comes from green plants. Sugar helps these plants grow. However, some green plants (like sugarcane and sugar beets) make more sugar than they need. The sugarcane stores the extra sugar in its long stalk. The sugar beet stores it in its fat root. Candy can be made with either cane sugar or beet sugar.

Chemistry is the study of chemicals and how they affect each other. You now know more about chemicals, especially the ones you've been testing, than you did before. There are lots more chemicals—everywhere. (Even you are made of chemicals!) Many are useful. Can you think of some useful chemicals you use every day?

Record Sheet 12-A

Name: _____

Date: _____

Chemical Information Sheet

1. I am made of crystals. When I am mixed with water, I dissolve. Some crystals may settle on the bottom, but the water is clear. When I am heated, I melt, turn golden brown and then black, and smell like caramel.

    I am the _____ unknown.

    I am *sugar*.

2. I fizz and bubble when vinegar is added to me. I do not burn when I am heated. When red cabbage juice is added to me, I turn green.

    I am the _____ unknown.

    I am *baking soda*.

3. I do not dissolve in water. When my water mixture is filtered, I am left in the filter paper. I turn purple-black when iodine is added to me. When I am heated, I turn brown.

    I am the _____ unknown.

    I am *cornstarch*.

4. I do not easily mix with water. When my water mixture is filtered, I am left in the filter paper. I have a pleasant smell. I do not burn when I am heated.

    I am the _____ unknown.

    I am *talc (baby powder)*.

5. When I am mixed with water, I dissolve. When the water mixture is left standing for a few days, I reappear as beautiful crystals. I turn bright purple when red cabbage juice is added to me.

    I am the _____ unknown.

    I am *alum*.

LESSON 13

Identifying the "Mystery Bag Chemical"

Overview and Objectives

In this lesson, students are challenged to apply their problem-solving skills to identify the mystery bag chemical from Lesson 1. To solve the problem, students begin to learn that they need to decide what tests they will perform and then compare these new test results with the known test results for the five chemicals they have been working with: sugar, baking soda, alum, cornstarch, and talc. As students begin to discover that they can use a variety of strategies to solve the problem, many may realize that it is possible to identify the contents of the mystery bag on the basis of one specific test.

■ Students review and discuss the processes they used to identify the unknown solids.

■ Students develop and apply a testing strategy to identify the contents of the mystery bag introduced in Lesson 1.

■ Using their test results, students analyze their recorded data, draw conclusions, and support those conclusions.

Background

Whether one is a scientist or detective, the process of solving any mystery (or problem solving) involves a combination of essential steps that include investigating, making observations, recording these observations or test results, analyzing the results, and drawing conclusions based on the results.

As students perform various chemical tests to identify the mystery bag chemical, they will discover that some test results reveal more information about the properties of the unknown chemical than others. For example, the results of the iodine test will quickly reveal a notable property of this chemical: it turns purple-black when iodine is added, thereby identifying it as cornstarch.

Some students may use other test results to support their conclusions. Or, they may identify the chemical incorrectly. Once again, keep in mind that getting the "right" answer is not as important as basing conclusions on actual test results.

Note: Because this is the first test in which students formally apply their skills to a new problem, they may need more of your guidance than in the past few tests. Lessons 14 and 16 also are "application" lessons, both of which will provide further opportunities for students to work independently.

Materials

For each student

  1   science notebook

  1   pencil

  1   **Record Sheet 13-A: Mystery Goo Test Results Table**

For every two students

  1   science pail

  5   toothpicks

  1   tray

  1   test mat

  1   sheet of wax paper, 16 x 22 cm (6 x 8½″)

  1   resealable plastic bag, 1 liter (1 qt), containing a sample of mystery goo (cornstarch and water)

  1   dry-erase marker

      Cleanup supplies

For the class

  1   resealable plastic bag, 1 liter (1 qt)

      Cornstarch, 120 ml (½ c)

      Water, 50 ml (2 oz)

      "How We Are Finding Out about the Unknowns" list (from Lesson 3)

Preparation

1. Copy **Record Sheet 13-A: Mystery Goo Test Results Table** for each student.

2. Make a new batch of mystery goo:

   ■ Put ½ cup of cornstarch into a plastic bag and add 50 ml (about 2 oz) of water.

   ■ Knead the bag to mix the goo thoroughly.

   ■ Divide the goo into 15 1-quart plastic bags. Add the bags, along with the dry-erase markers, to the materials center.

3. Flip over the test mats at the materials center so that the side with the unlabeled circles is visible.

4. Have student helpers cut and trim the 15 sheets of wax paper and add them to the materials center.

5. Check a dropper bottle of iodine solution from one of the science pails to see if it is still usable. Put a few drops on some of the goo. If the solution does not turn purple-black, the iodine has lost its indicator properties. In that case, empty and refill all 15 of the students' iodine dropper bottles with new iodine solution from the stock bottles.

6. Display the "How We Are Finding Out about the Unknowns" list.

Management Tip: You may want to pair students with new partners. This will enable you to evaluate each student's ability to apply his or her knowledge to a new situation and with a new partner.

Procedure

1. As a review, ask students to describe the steps they have taken to identify the five unknowns. Go over the "How We Are Finding Out about the Unknowns" list and ask students what else they need to add. In a class discussion, help students realize that important steps included testing or investigating, observing, recording their test results, analyzing these results, and consulting the Chemical Information Sheet.

2. Next, hold up a mystery bag of goo and remind the students that they observed it in Lesson 1. Let students know that the chemical mixed with water in the goo is, in fact, one of the five unknowns they have been working with. Ask what they might do to find out which one it is. Record their ideas on the chalkboard. If students do not suggest applying the tests they have performed on the five unknowns and using the information in their notebooks, suggest it.

3. Hand out **Record Sheet 13-A: Mystery Goo Test Results Table.** Go over it with the class. Explain that each team will get a sample of the goo. Their challenge is to use any of the materials in their science pails, as well as any of the information in their notebooks, to help discover the unknown's identity.

4. Have a student from each team pick up a mystery bag from the materials center. Remind students that in Lesson 1, they added water to the mystery chemical. Therefore, they don't need to perform a water test today. Ask students to observe the goo. Then ask them to write in their notebooks what they think the mystery goo might be. They should base their predictions on their observations and information in their notebooks.

5. Now review the **Student Instructions for Testing the Mystery Goo** on pg. 133 of this guide (pg. 59 of the Student Activity Book).

   Note: It is particularly important that students understand how to use **Record Sheet 13-A,** since they will use this format again in Lessons 14 and 16.

6. Answer any procedure-related questions. Then have teams pick up their materials and begin testing.

7. Circulate around the room. Check for the use of the compare circle. Talk with teams about which tests they are choosing.

8. Ask students to clean up. Make sure they throw away the goo bags and the wax paper.

Final Activities

1. Ask students to share their test strategies and results. The following questions may help to guide the discussion:

   - Which tests did you choose to do? Why?
   - Which test result gave you the most information about the properties of the unknown chemical? Why?
   - What information in your notebooks did you find the most useful?
   - How did your prediction about the identity of the goo compare with your results?

2. Ask a few students to share what they wrote in the "What I Think It Is" and "Why I Think So" sections of **Record Sheet 13-A.** If students support their conclusions with a negative result (for example, "it did not bubble with vinegar"), you may want to discuss the idea that negative results can be as important as positive results. (Students will explore the concept of negative results in Lesson 14.)

3. Collect **Record Sheet 13-A** for use in reviewing individual student work.

Extensions

1. Give each student a copy of the logic problem on pg. 136. When students have solved the problem, ask them to create new logic problems for their classmates to solve.

2. Link the process of solving a problem to that of solving a mystery in a book the class has read (see the **Bibliography** for ideas).

Student Instructions for Testing the Mystery Goo

1. Set up your materials. Use the side of the test mat with the unlabeled circles. Cover it with wax paper.

2. Using one of your measuring spoons, put a sample of the goo inside the compare circle.

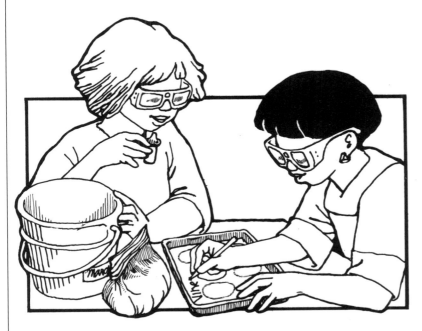

3. With your partner, decide which test you want to do first. Use the marker to write the test's name on the wax paper above one of the circles. Also record the test's name in the "What I Did" column of **Record Sheet 13-A.**

4. Put a sample of the goo in the circle you have just marked. Test the goo. Record your observations in the "What Happened" column of the record sheet.

5. Decide on another test. Then repeat Steps 3 and 4.

6. Continue testing and using the information in your notebooks until you think you know which of the five unknowns the goo contains. Fill in the "What I Think It Is" box on the record sheet. Then write in at least two reasons in the "Why I Think So" box.

Record Sheet 13-A

Name: _____

My Partner's Name: _____

Date: _____

Mystery Goo Test Results Table

What I Did	What Happened
What I Think It Is	**Why I Think So** **(Give two or more reasons)**

		Green Unknown	Yellow Unknown	Blue Unknown
Raoul's chemical dissolves in water and bubbles when vinegar is added.	Raoul			
Paula's chemical does not mix well with water and does not bubble when vinegar is added.	Paula			
Wendy's chemical dissolves in water and has no odor.	Wendy			

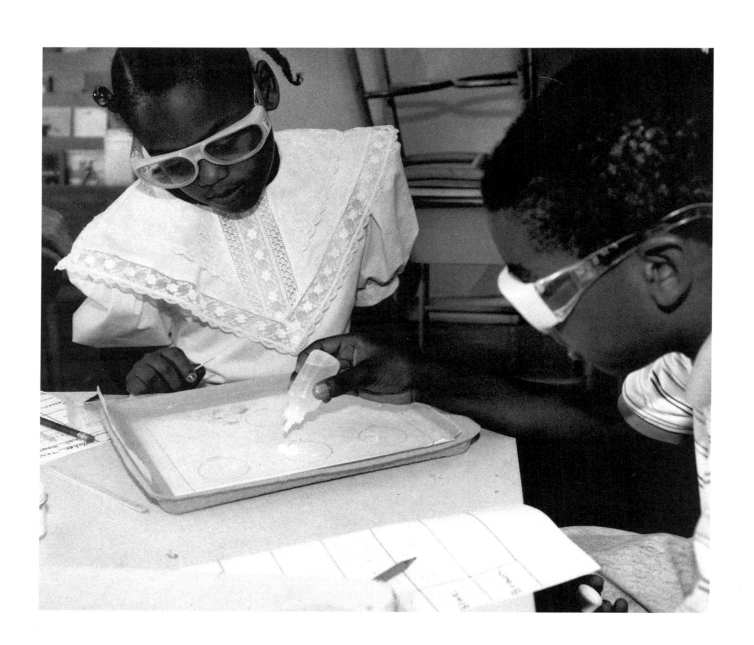

Testing Mixtures of Two Unknown Solids

Overview and Objectives

This lesson provides another opportunity for students to apply their laboratory skills and knowledge about the five chemical solids in solving a more difficult problem. In Lesson 13, students identified one solid; in this lesson they are challenged to identify two solids in a mixture. To do this, students first develop a plan that includes deciding which of the various tests they have learned will be an effective starting point. Some students may apply the deductive reasoning used in Lesson 13 to eliminate the need for performing all of the tests. Students also will find that they need to analyze both positive and negative results, as well as combinations of results, in order to identify the two solids in the mixture.

- Students decide which tests to perform, and in which order to perform them, to identify two solids in a mixture.

- Students interpret their test results, draw conclusions, and support those conclusions with their data.

Background

To prepare for this lesson, you will create three different mixtures, each containing two different chemicals.

- Mixture A: Cornstarch and baking soda

- Mixture B: Sugar and alum

- Mixture C: Baking soda and talc

One-third of the class receives mixture A; one-third, B; and one-third, C. Working in pairs, students decide which tests they want to perform and in what order.

When students apply the tests they have learned to a mixture of two of the five chemicals, each chemical reveals at least one of its main identifying properties (see Figure 14-1). Testing also reveals properties that are true for more than one chemical. For example, if the mixture does not mix well with water, it could contain either talc or cornstarch. In this case, students would need to perform another test to reveal a second property. For instance, if they smelled the mixture and determined it had no odor, they could rule out talc (baby powder) as a possibility and confirm that one of the chemicals is cornstarch. In this case, the absence of odor also illustrates the importance of negative test results—a concept the class will explore in greater depth in the **Final Activities.**

Figure 14-1

Identifying Properties Table

Chemical	Property
Sugar	Turns golden brown and smells like caramel when heated
Alum	Turns purple when red cabbage juice is added
Talc	Has a pleasant odor when observed with a sense of smell
Baking soda	Bubbles or fizzes when vinegar is added and turns green when red cabbage juice is added
Cornstarch	Turns purple-black when iodine is added

Management Tip: During this lesson, students will be performing a number of different tests simultaneously. Because the heat test requires close supervision to ensure safety in the classroom, it is not formally included in this lesson. If, however, you have another adult in the classroom who can oversee a heat station, you may wish to add this test as an option.

Materials

For each student

1 science notebook
1 **Record Sheet 14-A: Unknown Mixtures Test Results Table**
1 pencil

For every two students

1 science pail
1 tray
1 test mat
1 sheet of wax paper
5 toothpicks
1 labeled plastic soufflé cup and lid, 37 ml (1¼ oz), containing a sample of unknown mixture A, B, or C
1 dry-erase marker

For the class

1 large container in which to store optional student swap mixtures (see **Appendix A,** pg. 177)
1 measuring cup, 60 cc (¼ cup)
3 containers (e.g., resealable plastic bags, plastic tubs)
5 stock containers of the five unknowns
30 blank labels
30 plastic soufflé cups and lids, 37 ml (1¼ oz), for optional student swap activity
  Cleanup supplies

Preparation

1. You will be dividing the number of student pairs into thirds and giving each team one of the three unknown mixtures. Make them by mixing the following materials:

   ■ Mixture A: ¼ cup of cornstarch (orange unknown) and ¼ cup of baking soda (blue unknown)

   ■ Mixture B: ¼ cup of sugar (red unknown) and ¼ cup of alum (yellow unknown)

   ■ Mixture C: ¼ cup of baking soda (blue unknown) and ¼ cup of talc (green unknown)

   Fill five of the fifteen plastic soufflé cups with unknown mixture A, five with B, and five with C. Write the corresponding letters on blank labels and put them on the cups. Put on the lids. Add these to the materials center.

2. Have student helpers trim the 15 wax paper sheets and add them to the materials center.

3. Copy **Record Sheet 14-A: Unknown Mixtures Test Results Table** for each student.

4. Pair the students.

5. Read through final assessment 1, the student swap activity, in **Appendix A**. If you decide to do it, put blank labels on the remaining 30 plastic soufflé cups. Add these to the materials center. Write the steps for making the student swap mixtures on the chalkboard (see **Procedure,** Step 5).

Procedure

1. Hold up one cup of each unknown mixture. Explain that you have a new mystery for the class to solve. Then explain that each cup contains a different mixture of two of their five chemicals. Each team will get a cup containing one of the three mixtures.

2. Ask students, "What are some ways you might find out the identities of the two chemicals in these mixtures?" If necessary, remind students what they did in the last lesson.

3. Next, hand out and review **Record Sheet 14-A** with the class. Point out that the sheet uses the same format as **Record Sheet 13-A.** Each team should write the letter of the unknown mixture on the record sheet. Explain that teams will choose their own testing methods and the order in which to apply them, just as they did in Lesson 13. Briefly discuss possible testing methods (see Figure 14-2 for general test guidelines).

4. Provide the following information to the class:

   ■ Students will be trying to identify *two* chemicals in the mixture, as opposed to one.

   ■ Nothing has been added to the mixtures yet.

   ■ Students may choose any of the tests they have learned to help identify the chemicals. (If you decide to let students perform the heat test, make sure the testing is supervised by you or another adult.)

   ■ Students may use the information in their notebooks to help them identify the chemicals.

   ■ They may also refer to the instructions on pg. 59 in the Student Activity Book for testing reminders.

5. If your class will be doing final assessment 1 in **Appendix A,** explain that when students have finished testing, each will create an unknown mixture for a student swap activity at the end of the unit. Show students the instructions you have put on the chalkboard and review them with the class. To make the mixtures, each student needs to

    ■ Pick up a plastic soufflé cup and lid.

    ■ Use the measuring spoons to combine two or three measures of up to three of the chemicals.

    ■ Write his or her name on the cup's label.

    ■ Record in his or her notebook which chemicals were used.

    ■ Put the swap mixture at the materials center.

6. Have the teams pick up their materials. Then distribute the mystery mixtures. (Make sure teams near one another receive different mixtures.) Ask them to start testing (see Figure 14-2).

7. As students finish, ask them to clean up their areas. If they are doing the final assessment activity, remind them to take their swap mixtures to the materials center.

Final Activities

1. Have teams share their testing strategies, the reasoning behind them, and their conclusions. When students cite negative test results, use questions such as the following to elicit discussion:

    ■ How did negative test results help you identify the two chemicals in the mixture?

    ■ In what ways were the negative test results as helpful as the positive test results in identifying the chemicals ?

2. Collect **Record Sheet 14-A** for your review.

3. Ask the class for ideas about other people who might use a problem-solving method similar to the one the class has been using in this unit. List some of the students' ideas on the chalkboard. Other students' responses have included detectives, doctors, lawyers, and teachers.

Extensions

1. Invite another professional into your classroom (for example, a detective, a scientist, a mystery writer, an engineer, or another teacher) to discuss the nature of his or her work, especially as it relates to answering questions or problem solving. Before the visit, provide the individual with information on what students have been working on in this unit.

2. Have students review books such as *Messing Around with Baking Chemistry,* by Bernie Zubrowski, to choose activities they might like to explore at home or in classroom learning centers (see the **Bibliography** for references).

3. Copy the logic problem on pg. 147. Challenge your students to solve it. Ask the class how the logic problem was like the activities in this lesson.

Figure 14-2

*Testing the
unknown mixture*

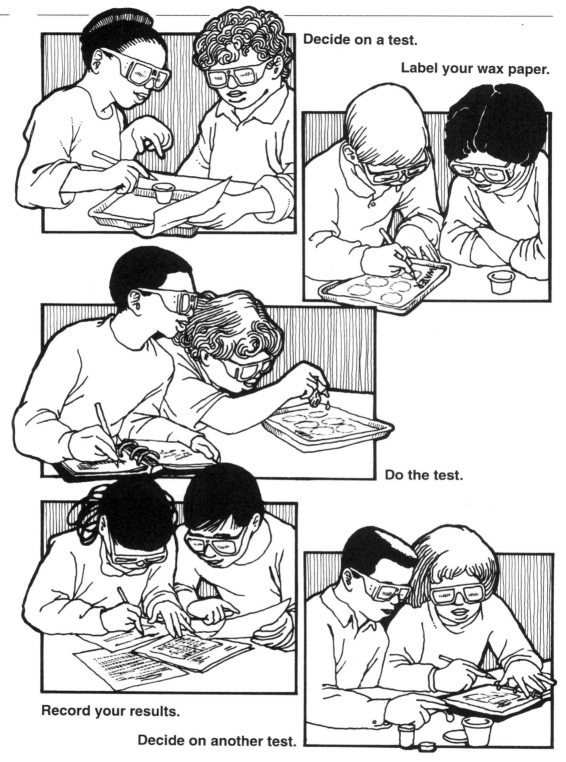

Decide on a test.

Label your wax paper.

Do the test.

Record your results.

Decide on another test.

Figure 14-3

Sample record
sheets

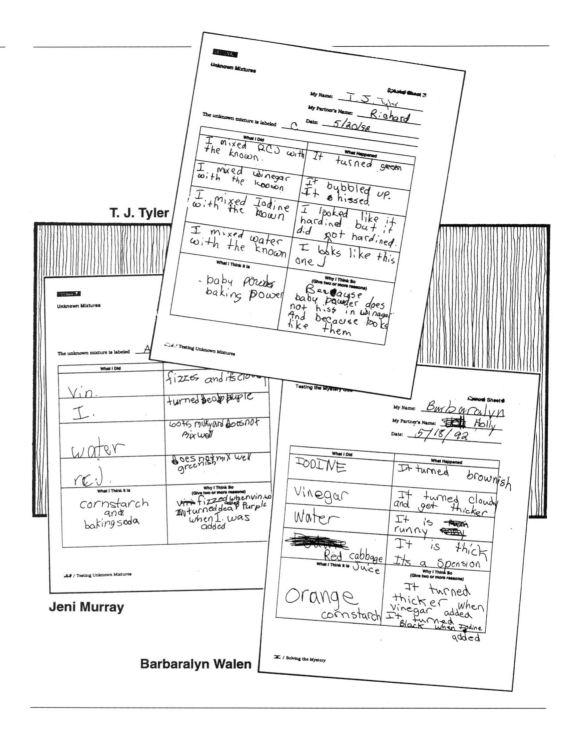

T. J. Tyler

Jeni Murray

Barbaralyn Walen

Assessment

As you review **Record Sheets 13-A** and **14-A** and listen to the class discussion in the **Final Activities,** consider the following questions:

■ Did students select the tests that would reveal the most information to help them solve their problems?

■ Did students perform a test, derive information from it, and apply that information in choosing the next test?

■ Do the test results students recorded show evidence of careful and complete observations?

■ Are the test results descriptive? Do they contain observable properties?

■ Do students give at least two reasons to support their conclusions? Are the reasons based on recorded test results?

Figure 14-3 shows examples of work products you can expect from students. As you evaluate the record sheets, use the questions above as criteria.

When you have finished reviewing the two record sheets, be sure to return them to students for later use.

Management Tip: If you have not recently checked the red cabbage juice in the science pails for its indicator properties, do so now. Students will use it in Lesson 15. Consider making a new batch (see **Appendix D** for directions) to ensure that your students have the best possible test results.

Record Sheet 14-A

Name: _____

My Partner's Name: _____

Date: _____

The unknown mixture is labeled _____

Unknown Mixtures Test Results Table

What I Did	What Happened

What I Think It Is	Why I Think So (Give two or more reasons)

		Baking Soda and Alum	Cornstarch and Alum	Sugar and Talc
Peter's mixture has a pleasant odor and turns golden brown and syrupy when heated.	Peter			
JenLee's mixture turns green and bright purple when red cabbage juice is added.	JenLee			
Max's mixture turns purple-black when iodine is added. When water is added and the mixture is filtered and left to evaporate, crystals appear.	Max			

Testing Household Liquids with Red Cabbage Juice

Overview and Objectives

In this unit, students have performed a series of tests that enabled them to identify unknown solids. In this lesson, students return to the red cabbage juice test from Lesson 9 and learn about the color changes they observed. By comparing test results on six new household liquids with information gleaned from a reading selection, students are introduced to a new concept: red cabbage juice is an indicator for three important groups of chemicals—acids, bases, and neutrals.

■ Students apply the red cabbage juice test from Lesson 9 to six household liquids.

■ Students share test results and develop their own system of classification for the six liquids.

■ Through a reading selection, students learn about acids, bases, and neutrals.

■ By comparing their test results with information in the reading selection, students classify the six household liquids (and their five powders) as acids, bases, or neutrals.

Background

As mentioned in Lesson 9, red cabbage juice, like many other plant materials, is a natural indicator—a substance that, through its color, reveals the presence of certain chemicals. Red cabbage juice is one indicator that can be used to determine whether substances are acids, bases, or neutrals.

Acids, bases, and neutrals are three kinds of chemicals. Acids (such as vinegar and lemon juice) usually are sour and react with many metals. Some are poisonous and can burn. Acids turn red cabbage juice a range of colors, from red-pink to deep purple. Acids are important components of fertilizers, polishes, soft drinks, and car batteries.

Most bases (such as baking soda and toothpaste) are bitter, feel slippery, and can dissolve fat and oil. They turn cabbage juice blue-green. Bases are found in some household cleaners, detergents, antacids, plasters, cements, and medicines.

Some substances, such as water, are neither acids nor bases. They are considered neutral. An acid solution will become neutral if you add a certain amount of base. Conversely, a base solution will become neutral if you add a certain amount of acid. (This process is called "neutralization.") When mixed with a neutral substance, red cabbage juice remains blue-purple. It undergoes no change.

In this lesson, students will add red cabbage juice to six household liquids. The test results your students are most likely to obtain appear in Figure 15-1.

Figure 15-1

Red Cabbage Juice Test Table

Household Liquid	Results	Acid, Base, or Neutral
Lemon juice	Turns pink	Acid
Detergent (washing soda)	Turns green	Base
Water	Stays color of juice	Neutral
Vinegar	Turns pink	Acid
Alcohol	Stays color of juice	Neutral
Ammonia	Turns green	Base

While the exact color of the red cabbage juice will vary from batch to batch, it always will be some shade of blue-purple. Make sure your class agrees on one color to describe the juice. What is important is that students observe this result: when each of the two neutrals (water and alcohol) is mixed with red cabbage juice, there is no color change.

When you ask the class to reach a consensus about the test results, it is important to remember that one student's perception of color may differ from another's (one student's "red" may be another's "pink").

To make sure that students are using the same color "vocabulary," you will ask them to hold up strips of colored paper to represent the color that they think most closely matches their test result. Assure students that when they finish the discussion, they will arrive at one set of results agreed upon by all.

When students disagree over results, take the opportunity to ask them what might be contributing to the variations in their test results and what students might do to resolve them. For example, other students have suggested repeating the activity, recording the most agreed-upon results, and explaining possible reasons for the varied results.

Management Tip: This is a long lesson. At this time, read through the **Final Activities** and decide if you want to do them the same day or the day after.

Safety Notes

- Household ammonia, used in a diluted 2% solution, is stored in small dropper bottles that will not spill. Because the solution is so dilute, it should not cause any irritation if it comes in contact with skin. If any irritation does occur, flush the skin surface with water. If students ingest it, give them two or three glasses of water followed by citrus fruit juice.

- Rubbing alcohol is used in a diluted 2% solution. Rubbing alcohol is toxic and is intended for external use only. If a student accidentally ingests alcohol, call your local poison control center immediately.

- If students accidentally get the dilute detergent solution in their eyes, immediately flush the eyes with a large quantity of water.

Materials

For every student
1   science notebook
1   pencil
1   **Record Sheet 15-A: Household Liquids Test Results Table**
3   strips of blue-purple, green, and pink construction paper, 5 x 11 cm
    (2 x 4½")

For every two students
1   science pail
1   sheet of wax paper, 16 x 22 cm (6 x 8½")
1   test mat
1   dry-erase marker
1   tray
2   paper towels
1   dropper bottle of red cabbage juice, 7 ml (¼ oz)

For every six students (three teams of two)
6   labeled dropper bottles of household liquids, 7 ml (¼ oz) (water, lemon
    juice, detergent solution, vinegar, ammonia solution, alcohol solution)
1   tray

For the class
1   sheet of newsprint
3   markers (green, blue-purple, pink)
1   plastic funnel
3   paper labels for class bulletin board: acid, base, and neutral
6   stock bottles of household liquids:
        Ammonia solution, 2%
        Lemon juice
        Water
        Alcohol solution, 2%
        Detergent solution, 2%
        Vinegar
    Cleanup supplies
    Thumbtacks

Preparation

1. Have student helpers trim the 15 wax paper sheets and add them to the
    materials center.

2. Put the household liquid labels on the 30 dropper bottles to make five sets
    of six liquids each. Fill the bottles and put each set on a tray, ready for
    distribution but separate from the materials center.

3. Collect the red cabbage juice dropper bottles from the students' science pails.
    Empty the bottles. If the second stock bottle of red cabbage juice has turned
    pink or green, adjust its color back to blue-purple or make a new batch of juice
    (see the **Preparation** section of Lesson 9 and **Appendix D** for instructions).
    Refill the dropper bottles and add them to the materials center.

4. Make a copy of **Record Sheet 15-A: Household Liquids Test Results Table**
    for each student.

5. Label the newsprint "Testing Household Liquids with Red Cabbage Juice." Make a table (modeled after **Record Sheet 15-A**) for recording class results. List on the chalkboard the six liquids students will be testing.

6. Team students in pairs and then put three teams together, so that there are six students in each group.

7. Have student helpers cut 30 strips each of green, blue-purple, and pink construction paper, approximately 5 x 11 cm (2 x 4½") per strip.

8. Select a class bulletin board area for displaying test results. Make three labels for the class bulletin board: acid, base, and neutral.

Procedure

1. Ask students what they have learned about red cabbage juice. Other students have discussed the color changes they have observed when using the juice. Explain that today the class will test some new household liquids with the juice to find out more about chemicals in general.

2. Show students the list of household liquids on the chalkboard. Ask what they think might happen if they add red cabbage juice to each liquid. Why do they think so? Have them record their ideas in their notebooks.

3. Distribute **Record Sheet 15-A.** Have students sit with their partners. Point out which three teams will share each set of liquids.

4. Then, go over the **Student Instructions for Testing Household Liquids with Red Cabbage Juice** on pg. 156 (pg. 68 of the Student Activity Book).

5. Have students pick up their materials. Place a tray of household liquids with each group of six students. Let them start testing.

6. Have students clean up.

7. Focus the students' attention on the newsprint you prepared earlier. Explain that all students will share their own results and then try to arrive at one set the whole class can agree on. Distribute three colored strips (green, blue-purple, and pink) to each student.

   ■ Starting with one of the household liquids, ask students to check **Record Sheet 15-A** and silently hold up the colored paper that most closely matches their result.

   ■ Using that same color of marker, record the most agreed-upon result on the class table.

   ■ Repeat this for the remaining five liquids.

   If students disagree a lot about color results, invite them to discuss why that may be.

8. Ask students to open their notebooks and draw three large circles on a clean sheet of paper. Explain that you would like each team to arrange the agreed-upon set of class results into groups. If they would like, they can name each group (see Figure 15-2). Then have the teams share their groupings and the reasons for them.

Figure 15-2

Sample grouping

Johnathon Strand
May 6, 1993

minesota
Alcohol
Water

Mrs. Pinky
Lemon J.
Vinegar

Ninja T.
Ammonia
Detergent

Final Activities

1. As a class, read "The Case of the Disappearing Stomachache" on pg. 158 (pg. 70 of the Student Activity Book).

2. Ask students what new information they have learned in this lesson about chemicals. Some simply will explain new concepts about acids and bases from the reading selection. Others may make the connection between acids, bases, and neutrals and the household liquids they have just tested.

   Now help students make that connection by asking questions such as the following:

   ■ What did you learn about chemicals by reading the story?

   ■ Which of the six household liquids have properties similar to those of the chemicals in the reading selection?

   ■ How are the properties similar?

   ■ What do you think the red cabbage juice test can tell you about the six liquids?

3. Have students look at the groupings in their notebooks. Then ask the following questions:

   ■ Would you change the way you've grouped these liquids? If so, how? Why?

   ■ How can you group these liquids into acids, bases, or neutrals?

4. Place the three labels (acid, base, and neutral) across the top of the class bulletin board. Ask the class to help you classify the household liquids as acid, base, or neutral.

5. Focusing on the acids (vinegar and lemon juice), discuss their properties (for example, sour, strong odor, turns red cabbage juice pink). Repeat this discussion for the bases and neutrals.

6. Ask students how they can use their new information to discover more about the five unknown chemicals they tested earlier. Have students look at their test summary tables and review the results of testing each unknown with red cabbage juice. Ask students to answer the following questions in their notebooks:

   ■ How did each of the original five unknowns react with red cabbage juice?

   ■ Using what you just learned about acids, bases, and neutrals, what can you now say about the original five unknowns?

   ■ Where would you put each unknown on the class board of acids, bases, and neutrals?

7. Ask students to share their discoveries about the five unknowns. Point out that since alum (the yellow unknown) is a weak acid, it does not turn pink-red but rather bright purple. This shows that the cabbage juice test can differentiate between weak acids and strong ones.

8. Ask the class to decide which group—acid, base, or neutral—each of the five chemicals belongs in. Add the unknowns to the class board. Then have students put their record sheets and test summary tables back in their notebooks.

Extensions

1. Now that students have read about acids, bases, and neutrals, have them test additional chemicals to discover that they, too, are acids, bases, or neutrals. Set up a classroom learning center and ask students to bring in samples of household products they would like to test with red cabbage juice. Some interesting products to test are shampoo, toothpaste, soap, juices, ginger ale, milk, soft drinks, and tea. Point out that some shampoos are advertised as pH balanced (neutral). Ask students to discover whether the shampoos are truly neutral.

2. Bring in some different materials that contain starch (such as pasta, bread, and white paper towels) and a few that do not. Have the students explore what the iodine test can tell them about these materials (if the material turns purple-black, it contains starch).

3. Make other natural acid/base indicators. Collect some red or purple flower petals (from flowers such as roses, petunias, and violets). Crush the petals, cover them with alcohol, and let the mixture sit for a few hours. (Or, boil the petals in spring water for a few minutes.) Test household chemicals with the new indicator.

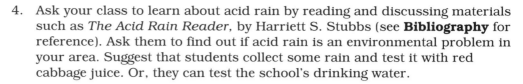

4. Ask your class to learn about acid rain by reading and discussing materials such as *The Acid Rain Reader,* by Harriett S. Stubbs (see **Bibliography** for reference). Ask them to find out if acid rain is an environmental problem in your area. Suggest that students collect some rain and test it with red cabbage juice. Or, they can test the school's drinking water.

5. Make vegetable art. Bring a red cabbage to class and cut it in half. Have students use pastel chalks to draw the cabbage's pattern. Then, have them add other patterned, colorful vegetables or fruits (such as kiwis, pomegranates, and strawberries) to their drawings.

Assessment

From this lesson, you can assess your students' abilities to analyze and interpret test results and to apply these results and information gleaned through reading to a new situation.

When reviewing student notebook entries and class discussions, consider the following criteria:

■ Are students able to draw conclusions based on their test results and information from a reading?

■ Do students demonstrate an understanding that the red cabbage juice test can indicate whether a substance is an acid, a base, or a neutral?

■ Were students able to apply what they learned as a result of testing the red cabbage juice with the household liquids to the test results of the five original unknown solids?

Student Instructions for Testing Household Liquids with Red Cabbage Juice

1. Set up your tray with the test mat and the wax paper over it.

2. Choose a household liquid to begin your test. Record its name on **Record Sheet 15-A.**

3. Draw a seventh circle on your test mat to use as your compare circle. Put three drops of the household liquid in the compare circle.

4. Using the dry-erase marker, write the household liquid's name above the first test circle. Put six drops of the liquid in the circle.

5. Add six drops of the red cabbage juice to the liquid you are testing.

6. Observe the mixture for about 15 seconds and record what happens on **Record Sheet 15-A.**

7. With a paper towel, remove the liquid from the compare circle.

8. Repeat Steps 2 through 7 for the five other household liquids.

Reading Selection

The Case of the Disappearing Stomachache

Acid or Base: How Can You Tell?

Has this ever happened to you? You take a big bite out of a sour pickle. You like it so much that you eat three more. Chances are, you get an awful stomachache. So, you take some stomach medicine. Soon, your stomach stops burning and you feel better. What's going on? Here's a clue. It has to do with two groups of chemicals: acids and bases.

Acids are found in foods like lemons, clear sodas, apples, vinegar, and, of course, pickles. Most of these acids taste sour and have a strong smell.

Then, there are the kinds of acids you **don't** eat, because they are poisonous. Some of these are used in fertilizers, polishes, and car batteries. Many of these acids are so strong that they can burn your skin or clothes.

Bases are found in detergent, oven cleaner, cement, baking soda, bleach, and the pills you take to make your stomach feel better. Some bases have a bitter taste and some burn. They're often slippery like soap.

How can you tell whether a chemical is an acid or a base? In the 1600s, a scientist named Robert Boyle did some experiments using the juices from plants such as violets and roses. When he added acids to the plant juices, they turned colors— either pink, red, or bright purple. When he added bases to the juices, they turned green.

We call these juices (like the red cabbage juice you used in your tests) **acid-base indicators**. Why? Because the juices indicate (or tell), by a change in color, whether a chemical is an acid or a base.

What about the chemicals that don't turn these plant juices pink or green? We call these substances **neutrals**. Neutrals—like water—are not acids or bases. But when you mix the right amounts of an acid and a base, you get a neutral substance. That process is called **neutralization**.

Why Did Your Stomachache Disappear?

So why did your stomachache go away? The pickles you ate caused too much acid to build up in your stomach. And the stomach medicine is a base.

When you swallowed the medicine, it mixed with the acid in your stomach and neutralized it. And you felt much better. (Next time a bee stings you, have an adult put some baking soda on it. What do you think will happen?)

Now, go back to your chart of results from the red cabbage juice tests.
What are the acids?
What are the bases?
What are the
neutrals?

Record Sheet 15-A

Name: _____

My Partner's Name: _____

Date: _____

Household Liquids Test Results Table

Name of Chemical	What Happened

Using the Known Solids to Identify Unknown Liquids

Overview and Objectives

In this final lesson, students encounter an interesting reversal: instead of using the test liquids—water, iodine, vinegar, and red cabbage juice—to identify the solids, they are challenged to use the properties of the solids to identify unlabeled test liquids. To solve this new problem, students must apply the problem-solving and chemical-testing skills they have developed throughout the unit. Through class discussion of their results, students will recognize that more than one strategy can be used to solve a problem and that some strategies require applying more tests than others. In addition, this culminating lesson enables you and your students to assess growth in two areas: the application of skills and the understanding of the concepts included in this unit.

■ Students decide which chemical tests they will perform and in what order they will perform them to solve a new problem.

■ Students analyze their recorded data, draw conclusions, and support these conclusions with their test results.

■ Students record their thoughts about the significance of negative results and about chemical properties as indicators.

Background

This activity reverses the testing procedure used in previous lessons. The chemical properties of the five unknowns now serve as indicators for three liquids in "disguise":

■ Liquid A is white vinegar.

■ Liquid B is water.

■ Liquid C is iodine.

To keep the class from identifying the liquids by their color, you will dye them with food coloring.

Students may discover that

■ Cornstarch is the indicator for iodine.

■ Baking soda is the indicator for vinegar.

■ All five chemicals are indicators for water.

Once again, remember that negative results are as valuable as positive ones. For example, if the mixture of baking soda and the unknown liquid does not bubble, you will know that the liquid is not vinegar.

Materials

For each student
- 1   science notebook
- 1   **Record Sheet 16-A: Unknown Liquids Test Results Table**

For every two students
- 1   science pail
- 1   tray
- 1   test mat
- 5   toothpicks
- 1   dropper bottle of unknown solution, 7 ml (¼ oz), labeled A, B, or C
- 1   dry-erase marker
- 1   sheet of wax paper, 16 x 22 cm (6 x 8½")

For the class
- 1   stock bottle of white vinegar, 250 ml (½ pt)
- 1   stock bottle of 0.1% iodine solution, 250 ml (½ pt)
- 1   package of food coloring
- 1   plastic funnel
- 3   plastic cups, 207 ml (7 oz), for disguising unknowns
- 3   dropper bottles, 7 ml (¼ oz)
- 3   blank labels
-      Cleanup supplies

Preparation

1. Have student helpers trim the 15 wax paper sheets and add them to the materials center.

2. To make the iodine, vinegar, and water look similar, add food coloring to 50 ml of each liquid. Since the iodine is already colored, you can try to match its shade by adding combinations of red and yellow food coloring to equal amounts of water and vinegar. Or, you can make each liquid a different color. The goal is to mask the property of color so that students cannot easily identify the liquids by their colors.

3. Fill five dropper bottles with each unknown liquid:

   - Unknown Liquid A: White vinegar

   - Unknown Liquid B: Water

   - Unknown Liquid C: Iodine solution

   Write the corresponding letters on each blank label and put the labels on the dropper bottles. Have the unknown liquids ready to distribute.

4. Copy **Record Sheet 16-A: Unknown Liquids Test Results Table** for each student.

5. Pair the students.

6. In the **Final Activities** you can choose whether to reveal the identities of the three unknown liquids or to do Extensions 1 or 2 on pg. 167. Read through the extensions now to decide.

Procedure

1. Ask the class to recall what liquids they used in their chemical tests (water, vinegar, iodine, and red cabbage juice). Hold up bottles A, B, and C, and explain that each contains one of the liquids the students used to test the five unknowns. Explain that the liquids have been dyed so that they cannot be identified by their property of color.

2. Ask students to suggest some ways they could identify the unknown liquids. Remind them that they can use their notebooks, Chemical Information Sheets, and test summary tables to help solve this new mystery.

3. Hand out **Record Sheet 16-A.** As you review it with the class, point out where to record the letter of the unknown liquid being tested. Explain that they will choose their own testing methods.

4. By now, students know the testing procedures well. Explain that students should use one of the six circles as a compare circle in which to put a few drops of the unknown liquid.

5. Have the students pick up their materials.

Figure 16-1

Testing unknown liquids

6. Give one bottle of unknown liquid to each team. Make sure teams near one another receive different liquids.

7. Circulate as the teams work, checking for skills such as those listed in the **Assessment** section on pg. 167. As teams finish, ask them to clean up.

Final Activities

1. Ask students to record in their notebooks answers to the following questions:
   - Which tests gave you the most information?
   - Which properties helped you identify the unknown liquid? Why do you think these properties were so helpful?
   - How did you use negative results to support your conclusions?

2. Have teams share information on their testing procedures, results, conclusions, and supporting data.

3. Collect the record sheets.

4. Reveal the identities of liquids A, B, and C (unless you are going to do Extensions 1 or 2).

Figure 16-2

Samples of Record Sheet 16-A

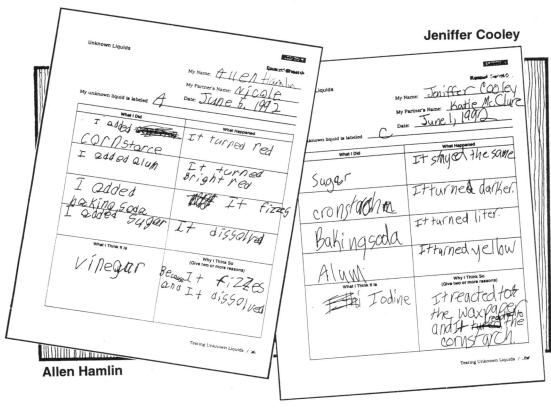

Extensions

1. Have students test the other unknown liquids, including the red cabbage juice. Then reveal their identities.

2. Switch the labels on the unknown liquids. Add a drop of perfume to each to remove odor, which is another identifying property. Let the students test these unknown liquids.

3. Have the class create a "Chemical Information Sheet" (like the one used to identify the five unknowns in Lesson 12) for the unknown liquids. Ask groups of students to write "What Am I?" descriptions using the properties of each liquid. You can also include the red cabbage juice.

Assessment

This lesson offers you several opportunities to assess students' current ability to apply tests, interpret information, and relate that information to aspects of chemistry. Because this is the end of the unit, you also will want to examine the post-unit assessment on pg. 171. In addition, final assessment activities are offered in **Appendix A.**

In this lesson, students' testing strategies, discussions, notebook entries, and record sheets will help you assess growth in the following areas:

- Students' ability to record observations and experiences that are informative and are not based on inference or unobservable properties.

- Students' ability and willingness to verbalize what they are doing, what they have discovered, and what they can conclude.

- Students' ability to devise testing strategies and solve problems through the application of new knowledge and skills.

- Students' ability to perform the physical and chemical tests in this unit.

- Students' skill and confidence in handling materials.

- Students' ability to analyze test results and draw conclusions based on them.

- Students' ability to support their conclusions with reasons based on experiences.

- Students' ability to work cooperatively.

- Students' ability to use new vocabulary appropriately.

Note: You may now want to return the record sheets you have collected. Students can take them home (along with the rest of their notebooks) when the unit ends.

Post-Unit Assessment

The post-unit assessment is a matched follow-up to the pre-unit assessment in Lesson 1. By comparing students' pre- and post-unit responses, you will be able to document their growth in knowledge about the properties of chemicals.

Final Assessments

The final assessments in **Appendix A** include a self-assessment for students and an activity in which students analyze the composition of unknown mixtures.

Management Tip: Students will need their science pails for the post-unit assessment and final assessments. After that, have the class empty and clean their five unknown jars. Also have students clean the measuring spoons and goggles before putting them away.

Record Sheet 16-A

Name: _____

My Partner's Name: _____

Date: _____

The unknown liquid is labeled _____

Unknown Liquids Test Results Table

What I Did	What Happened
What I Think It Is	**Why I Think So** (Give two or more reasons)

Post-Unit Assessment

Overview

This is the second part of the matched pre- and post-unit assessments of students' ideas about chemicals. By comparing the individual and class responses from Lesson 1 with those from the following three activities, you will be able to document students' learning over the course of the unit. During the first lesson, students

■ Individually recorded their thoughts and questions about chemicals

■ Developed class lists entitled "What We Think about Chemicals" and "What We Would Like to Know about Chemicals"

■ Individually wrote about the "unlabeled container" scenario

When they revisit these activities during the assessment, students are likely to appreciate how much they have learned about household chemicals and the ways in which chemicals interact.

Materials

For each student
1   science notebook

For the class
Newsprint
Markers
Lists from Lesson 1: "What We Think about Chemicals" and "What We Would Like to Know about Chemicals"

Procedure

Student Notebook Entries

1. Explain that the purpose of this assessment is to give students a chance to review the materials in their notebooks in order to help them recognize what they have learned.

2. If you wish, divide the students in groups of four and have them use a "jigsaw" approach to share thoughts. Ask each group to review a particular activity or series of activities, share what they have learned among themselves, and then share with the whole class. Finally, ask the class to add any comments.

3. Have students write a final notebook entry describing what they now know about chemicals and chemistry. Also ask them to include any questions they still have or any new ones that have arisen since they began to work on this unit.

Figure PUA-1

Sample pre- and post-unit notebook entries

Wendy Ellis

Class Lists

1. Display the newsprint and title it "What We Think about Chemicals." Conduct a class brainstorming session and record students' thoughts.

   Focusing on the earlier list, ask students the following questions:

   ■ What statements do you now know, without a doubt, to be true?

   ■ What evidence, such as experiences from this unit, can you offer to support these statements?

3. Ask students to correct or improve the statements that they think need revising and to give reasons for the changes. Then ask what statements are on the new list that do not appear on the list from Lesson 1.

4. Display the list entitled "What We Would Like to Know about Chemicals" from Lesson 1. Ask the following questions:

   ■ Which of these questions can you now answer?

   ■ What are some ways to find out the answers to questions that were not answered?

5. Encourage the class to continue looking for answers to questions that were not answered in the unit.

Sample of pre- and post-unit class lists

WHAT WE THINK ABOUT CHEMICALS Feb. 1

They can change things.
They are a liquid or a solid.
They are big and mostly small.
They come from different things.
They are liquid, flour, baking soda, powder, sugar and soda.
It's a liquid or a solid which scientists use, mainly archaeologists.
They might make you sick.
It could kill you.
It has acid in it. It can be liquid or powder.
It can be non-sticky.
They come from liquids or solids.
It is a wetness or a dryness.

WHAT WE THINK ABOUT CHEMICALS May 5

1. Some turn different colors when you add another chemical. It's called a chemical reaction.
2. Some chemicals can make other chemicals change color. (indicators)
3. Some are solutions and some are suspensions.
4. Some chemicals are powder (solid), some are liquid, some are gases.
5. Some chemicals evaporated and some got filtered.
6. Sometimes chemicals react with the wax paper, like the iodine, and changes the result.
7. People get different results sometimes.
8. Every thing has chemicals.
9. Negative results can help you.
10. You can do all sorts of tests on chemicals and then use all the information you've gathered to decide what it is. You can make a test chart.

6. When reviewing and comparing the pre- and post-unit lists, look for statements that reflect some of the following concepts:

- Are students more aware that chemicals are all around us?

- Have students developed an understanding that different chemicals have different properties?

- Are students more aware that changes occur when different materials are mixed, separated, or heated?

- Are students aware that there are different types of mixtures with different properties?

"Unlabeled Container" Scenario

1. Write the following scenario from Lesson 1 on the chalkboard: A label has peeled off a container of white powder in your kitchen. What are some ways you can find out what the powder is?

2. As you did in Lesson 1, ask students to write answers to the question.

3. Compare and contrast these answers with students' notebook entries from Lesson 1. As you review both sets of responses, consider the following questions:

- Is the student reflecting on and applying his or her experiences in the unit?

- Does the student's proposal for solving the problem indicate the application of previously learned knowledge and skills?

Figure PUA-3

Sample pre- and post-unit notebook entries

Date April 15, 1992

I think it is blue?

I would add water@see if it would stick to something.

Date

Name: Shawn   Date: June 9, 1992

Mystery white chemical

I would do lots of test and write gather all of the information and look for some other chemical that looks some what the same and do made the same thing I did to the chemical that had No lable.

Shawn Thomas

These activities provide you with a written record of each student's growth during this unit. Let those students who have difficulty recording their thoughts on paper share them orally. Refer to *Chemical Tests*: **Goals and Assessment Strategies** on pg. 10 to evaluate students' growth and understanding.

Final Assessments

Overview

Following are some suggestions for assessment activities. Although it is not essential to do both these activities, it is recommended that students do Assessment 2.

■ Assessment 1 is a student swap activity in which students analyze the composition of the unknown mixtures they created in Lesson 14.

■ Assessment 2 is a self-assessment that students can use to monitor their own learning.

Assessment 1

An Analysis of Unknown Mixtures from the Student Swap Activity in Lesson 14

Materials

For each student

 1   science notebook
 1   **Record Sheet A-1: Unknown Mixtures Test Results Table** (see pg. 181)
 1   pencil
 1   science pail
 1   tray
 1   test mat
 1   sheet of wax paper, 16 x 22 cm (6 x 8½″)
 5   toothpicks
 1   student swap jar
 1   dry-erase marker
     Cleanup supplies

Note: Because this assessment requires many materials, you may wish to have students do it in two groups of 15.

Procedure

1. Hand out a copy of **Record Sheet A-1: Unknown Mixtures Test Results Table** (pg. 181).

2. Review the record sheet. Explain that students may refer to their notebooks for information.

3. Have the first group of students pick up materials.

4. Distribute the student swap mixtures. Make sure students do not receive the mixture they created. Remind them not to discuss their results with others. Try to observe the students as they work.

5. After students have finished testing, have them clean up.

6. Collect **Record Sheet A-1.**

7. Repeat this procedure with the other half of the class.

8. Have teams share with the class which of the five chemicals were in the mixtures they created.

9. Review students' record sheets. Keep the following criteria in mind:

   ■ Does the student demonstrate skill in handling the materials?

   ■ Does the student know how to conduct the various physical and chemical tests?

   ■ Do the recorded test results include clear, descriptive observations?

   ■ Did the students select the tests that would reveal the greatest possible amount of information needed to help solve their problem?

   ■ Did the students perform a test, derive information from it, and apply that information in choosing the next test?

   ■ Can the student draw valid conclusions from the results?

   ■ Can the student provide reasons based on experiences to support his or her conclusions?

Remember that the above criteria, not whether the student correctly determines the mixture's composition, are most important when evaluating record sheets. Figure A-1 illustrates the variety of responses you may expect.

Assessment 2

A Student Self-Assessment

This assessment encourages students to examine their own progress. If possible, have students fill it out two times: after Lesson 10 and at the end of the unit. In that way, students can compare earlier and later responses and get a picture of their own growth.

Teachers have found it useful to meet with each student individually to discuss these self-assessments. Such meetings give you the opportunity to provide your feedback about the student's work and to compare it with the student's perception.

Figure A-1

Sample record
sheets

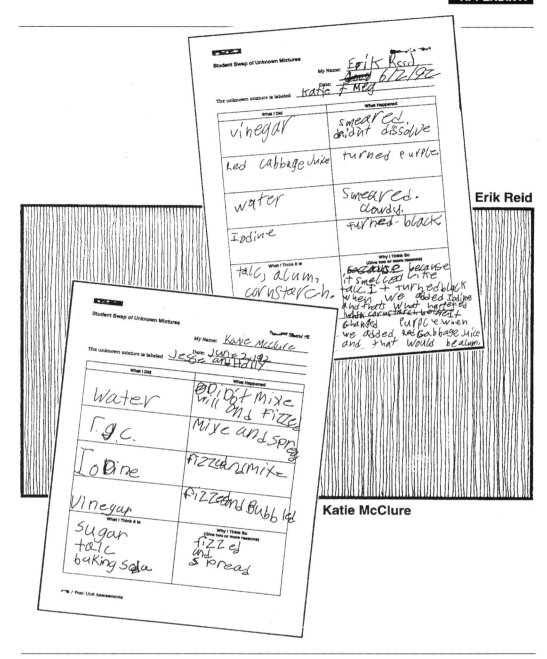

Erik Reid

Katie McClure

Materials

For each student

    1   Student Self-Assessment (see pg. 180)

Procedure

1. Distribute a copy of the Student Self-Assessment to each student. Review it with the class. Explain that from time to time it is important to stop and think about how we are working.

2. Allow students enough time to complete the assessment in class, or ask them to do it as a homework assignment.

Chemical Tests:
Student Self-Assessment

Name: _____

Date: _____

1. Write down two or three important things you have learned from doing the *Chemical Tests* unit.

2. How did you feel about working with the chemical testing materials in the unit? Did your feelings about the materials change as you worked through the unit? If so, give some examples.

3. Write down some activities in the unit you enjoyed. Explain why you liked them.

4. Were there any activities in the unit that you didn't understand or that confused you? Which ones? Why were they confusing?

5. Take another look at your record sheets and your science notebooks. Describe how well you think you recorded your observations and ideas.

6. How well do you think you and your partners worked together? Give some examples.

7. How do you feel about science now? Circle the words that apply to you.

   a) Interested   b) Nervous   c) Excited   d) Bored

   e) Confused   f) Successful

   g) Write down one word of your own _____

Record Sheet A-1

Name: _____

Date: _____

The unknown mixture is labeled _____

Unknown Mixtures Test Results Table

What I Did	What Happened

What I Think It Is	Why I Think So (Give two or more reasons)

APPENDIX B

Bibliography

The **Bibliography** is divided into the following categories:

- Resources for Teachers
- Resources for Teachers and Students
- Science Activity Books
- Related Trade Books
- Computer Programs
- Mystery Books

While not a complete list of the many books on chemistry, this bibliography is a sampling of resources that complement this unit. They have been favorably reviewed, and teachers have found them useful.

If a book goes out of print or if you seek additional titles, you may wish to consult the following resources.

Appraisal: Science Books for Young People (The Children's Science Book Review Committee, Boston).

> Published quarterly, this periodical reviews new science books available for young people. Each book is reviewed by a librarian and by a scientist. The Children's Science Book Review Committee is sponsored by the Science Education Department of Boston University's School of Education and the New England Roundtable of Children's Librarians.

National Science Resources Center. *Science for Children: Resources for Teachers.* Washington, DC: National Academy Press, 1988.

> This volume provides a wealth of information about resources for hands-on science programs. It describes science curriculum materials, supplementary materials (science activity books, books on teaching science, reference books, and magazines), museum programs, and elementary science curriculum projects.

Science and Children (National Science Teachers Association, Washington, DC).

> Each March, this monthly periodical provides an annotated bibliography of outstanding children's science trade books primarily for pre-kindergarten through eighth-grade science teachers.

Science Books & Films (American Association for the Advancement of Science, Washington, DC).

Published nine times a year, this periodical offers critical reviews of a wide range of science materials, from books to audiovisual materials to electronic resources. The reviews are primarily written by scientists and science educators. *Science Books & Films* is useful for librarians, media specialists, curriculum supervisors, science teachers, and others responsible for recommending and purchasing scientific materials.

Scientific American (Scientific American, Inc., New York).

Each December, Philip and Phylis Morrison compile and review a selection of outstanding new science books for children.

Sosa, Maria, and Shirley Malcom, eds. *Science Books & Films: Best Books for Children, 1988-91.* Washington, DC: American Association for the Advancement of Science Press, 1992.

This volume, part of a continuing series, is a compilation of the most highly rated science books that have been reviewed recently in the periodical *Science Books & Films*.

Resources for Teachers

Chisholm, Jane, and Mary Johnson. *Introduction to Chemistry.* London: Usborne Publishing, 1983.

A general introduction to chemistry that provides information on chemical reactions and chemical classification.

Dishon, Dee, and Pat Wilson O'Leary. *A Guidebook for Cooperative Learning: Techniques for Creating More Effective Schools.* Holmes Beach, FL: Learning Publications, 1984.

A practical guide for teachers who are embarking on the implementation of cooperative-learning techniques in the classroom.

Headstrom, Richard. *Adventures with a Hand Lens.* New York: Dover Publications, Inc., 1962.

Common objects are viewed from a new perspective. Simple line drawings and explanations make this an excellent resource.

Johnson, David W., Roger T. Johnson, and Edythe Johnson Holubec. *Circles of Learning.* Alexandria, VA: Association for Supervision and Curriculum Development, 1984.

This book presents the case for cooperative learning in a concise and readable form. It reviews the research, outlines implementation strategies, defines the skills students need to work cooperatively, and answers many questions.

National Chemistry Week, sponsored by the American Chemical Society, 1155 16th St., NW, Washington, DC 20036. Telephone: 1-800-227-5558.

The American Chemical Society designates one week every year to reach out to the public with positive messages about chemistry. Scientists are available to visit classrooms, and related curriculum materials are available free of charge.

Norton, Donna E. *The Effective Teaching of Language Arts.* Columbus, OH: Merrill Publishing Co., 1989.

> This book includes information on the use of semantic mapping and webbing.

Sattler, Helen Roney. *Recipes for Art and Craft Materials.* New York: Lothrop, Lee, and Shepard, 1987.

> Contains a collection of inventive recipes for making inexpensive materials for art and craft projects.

Resources for Teachers and Students

Billings, Charlene. *Microchip: Small Wonder.* New York: Putnam, 1984.

> Good book for students interested in finding out more about silicon chips.

Kramer, Stephen P. *How to Think Like a Scientist.* New York: Thomas Y. Crowell, 1987.

> How can we find the answers to questions, and how we can be sure the answers are correct? This book highlights the scientific method.

Stubbs, Harriett S. *The Acid Rain Reader.* Raleigh, NC: The Acid Rain Foundation, 1989.

> This booklet discusses the composition and major causes of acid rain.

Symes, R. F., and R. R. Harding. *Eyewitness Books: Crystals and Gems.* New York: Alfred A. Knopf, 1991.

> This book describes the seven basic shapes of crystals and how they form in nature. It also covers how crystals are studied and identified, grown artificially, and used in industry.

Whyman, Kathryn. *Chemical Changes.* New York: Gloucester Press, 1986.

> Discusses the states of matter, elements and compounds, and how chemicals can be mixed together to make the substances we use every day.

Science Activity Books

Cobb, Vicki. *Gobs of Goo.* New York: J. B. Lippincott, 1983.

> This book describes various types of sticky substances and shows how they are made and used in everyday life. The experiments are easy to do.

————. *Science Experiments You Can Eat.* New York: Harper and Row, 1972.

> A book of kitchen experiments with food. It demonstrates various scientific principles while producing eatable results and includes information about solutions, suspensions, and crystals. Other Vicki Cobb books to try are *The Secret Life of School Supplies* and *The Secret Life of Cosmetics.*

Gardner, Robert. *Kitchen Chemistry.* New York: Julian Messner, 1982.

> This book describes simple experiments to do in the kitchen laboratory.

Jennings, Terry. *Everyday Chemicals.* Chicago: Children's Press, 1989.

> An introduction to the many kinds of chemicals, this book describes some of their uses. It includes study questions, activities, and experiments.

Johnson, Mary. *Chemistry Experiments.* London: Usborne Publishing, 1981.

> This book is full of safe and simple experiments using equipment and chemicals found at home.

Maki, Chu. *Snowflakes, Sugar, and Salt: Crystals Up Close.* Minneapolis: Lerner, 1993.

> This book from the "Science All Around You" series examines the crystals of sugar, salt, alum, baking soda, and snowflakes, all of which are illustrated with beautiful photographs. Simple crystal-growing activities also are included.

Mebane, Robert C., and Thomas R. Rybolt. *Adventures with Atoms and Molecules (Books 1 and 2).* Hillside, NJ: Enslow Publishers, 1987.

> These books contain a variety of experiments and activities students can do at home and school. They demonstrate the properties and behavior of various kinds of atoms and molecules.

Robinson, Marlene M. *The Crystal Kit.* Philadelphia: Running Press, 1988.

> This kit contains the book *Crystals: What They Are and How to Grow Them,* a package of crystal mix (alum), and directions for simple crystal experiments.

Shalit, Nathan. *Cup and Saucer Chemistry.* New York: Dover Publications, 1972.

> This book contains simple experiments using materials available at home to illustrate basic principles of chemistry.

Van Cleave, Janice Pratt. *Chemistry for Every Kid.* New York: John Wiley and Sons, Inc., 1989.

> This book contains more than 100 chemistry experiments, each of which shows how chemistry is part of our everyday lives.

Wyler, Rose. *Science Fun with a Homemade Chemistry Set.* Old Tappan, NJ: Julian Messner, 1987.

> A good source of chemistry activities students can do at home with everyday chemicals.

Zubrowski, Bernie. *Messing Around with Baking Chemistry.* Boston: Little, Brown and Co., 1981.

> This museum activity book enables students to explore what happens when batter and dough turn into cake and bread. It emphasizes the properties of baking powder, baking soda, and yeast.

Related Trade Books

Cole, Joanna. *The Magic School Bus at the Waterworks.* New York: Scholastic, 1986.

This humorous book follows Mrs. Fizzle and her class on a field trip through the water cycle and a water filtration plant.

DePaola, Tomie. *Strega Nona's Magic Lessons.* New York: Harcourt Brace Jovanovich, 1982.

Bambalona is tired of working for her father, the baker, and sets off to learn magic from Strega Nona. Teachers can relate this story to chemical and physical changes.

———. *The Legend of the Indian Paintbrush.* New York: G. P. Putnam's Sons, 1988.

The story of a Plains Indian boy who follows his destiny to become an artist, this book offers one example of how plants are used for dyes.

Grey, Vivian. *The Chemist Who Lost His Head.* New York: Coward-McCann, Inc., 1982.

Recounts the life of the French chemist Antoine Lavoisier, whose work helped transform many of the undocumented scientific beliefs of the Middle Ages into an exact science.

Sabin, Louis. *Marie Curie.* Mahwah, NJ: Troll, 1985.

The biography of the Polish-born scientist who, with her husband, was awarded the 1903 Nobel Prize for discovering radium.

Shannon, George. *Stories to Solve: Folktales from Around the World.* New York: Greenwillow, 1985.

Fourteen folktales from around the world, each with a mystery or problem for the characters to solve.

Van Allsburg, Chris. *Two Bad Ants.* Boston: Houghton Mifflin Co., 1988.

Two bad ants desert their colony in search of marvelous crystals.

Computer Programs

MECC Mystery Objects. Available from MECC, 6160 Summit Dr. W., Minneapolis, MN 55430.

Children identify mystery objects by learning about their physical properties. Recommended by MECC for grades 2 through 4.

MECC Mystery Matter. Available from MECC, 6160 Summit Dr., W., Minneapolis, MN 55430.

This program has children explore the special properties of chemicals. It is divided into two parts. In Matter Search, children identify a piece of matter by testing for its special properties. In Matter Maker, children look at each chemical and create their own mystery matter for others to look at. Recommended by MECC for grades 3 through 6.

National Geographic Kids Network: Acid Rain Kit. Available from the
National Geographic Society, Educational Services, Dept. 5397,
Washington, DC 20036.

Students explore acid rain through research and experiments. After
designing and building their own rain collectors, they measure the acidity
of local rainwater. Classes compare their measurements with those taken
by other students in different parts of the world. Recommended for
grades 4 through 6.

Mystery Books

Good children's mystery tales not only convey an atmosphere of excitement and
suspense but also help students develop logical thinking. As a complement to
this unit, following are some mystery book series and single titles students can
read either independently or in read-aloud sessions. They are targeted at second-
through fourth-grade reading levels. While this list is a good starting point, it is
by no means a complete guide to the many wonderful books available.

If you want to help students link their experiences with chemical tests to the story
line of the mystery tales, see the Mystery Book Review record sheet on pgs. 190–91.

Adler, David A. "Cam Jansen" series. New York: Viking Press, 1980.

Bonsall, Crosby. *The Case of the Cat's Meow.* New York: HarperCollins
Children's Books, 1978.

Bulla, Clyde Robert. *The Ghost of Windy Hill.* New York: Scholastic, 1968.

Cameron, Eleanor. *The Terrible Churnadryne.* New York: Little, Brown
and Co., 1959.

Giff, Patricia Reilly. "The Polka Dot Private Eye" series. New York: Dell Young
Yearling, 1984.

Hope, Laura. *Mystery at School.* New York: Putnam, 1989.

Howe, James. "Bunnicula" series. New York: Macmillan, 1979.

Hutchins, Pat. *The Mona Lisa Mystery.* New York: Greenwillow, 1983.

Kastner, Erich. *Emil and the Detectives.* New York: Doubleday, 1930.

Ladd, Elizabeth. *A Mystery for Meg.* New York: Morrow, 1962.

Levy, Elizabeth. "Something Queer" series. New York: Dell Young Yearling, 1982.

Lexau, Joan. *The Homework Caper.* New York: HarperCollins Children's
Books, 1966.

———. *The Rooftop Mystery.* New York: HarperCollins Children's Books, 1968.

McArthur, Nancy. *The Adventure of the Buried Treasure.* New York:
Scholastic, 1990.

Meyers, Susan. *P. J. Clover, Private Eye: The Case of the Halloween Hoot.*
New York: Dutton, 1990.

Parrish, Peggy. *Clues in the Woods.* New York: Macmillan, 1968.

Pearson, Susan. "Eagle-Eye Ernie" series. New York: Simon and Schuster, 1990.

Sharmat, Marjorie Weinman. "Nate the Great" series. New York: Dell Young
Yearling, 1972.

Simon, Seymour. "Einstein Anderson" series. New York: Puffin Books, 1981.

Sobol, Donald J. "Encyclopedia Brown" series. New York: Bantam, 1966.

———. *Two Minute Mysteries.* New York: Bantam, 1986.

Warner, Gertrude Chandler. "The Boxcar Children" series. Morton Grove, IL: Albert Whitman and Co., 1989.

Ziefert, Harriet. *Mr. Rose's Class Mystery Day.* New York: Bantam, 1989.

Name: _____

Date: _____

Mystery Book Review

1. The book I am reading is _____

2. The mystery or problem to solve is _____
   _____
   _____

3. List the clues that help the characters solve the mystery or problem.
   _____
   _____
   _____
   _____
   _____

4. In the list above, check the clues you think were most helpful in solving the mystery.

Mystery Book Review *(continued)*

5. What was the solution to the mystery? _____

_____

_____

_____

_____

6. What did you think of this book? Would you recommend it to your friends? _____

_____

_____

_____

_____

_____

"Check Your Science Pail" Poster Assembly and Use

The "Check Your Science Pail" poster is an important materials management tool that places responsibility with the student. Throughout the unit, students refer to this poster to remind themselves what materials should be stored in their science pails at the end of each lesson. The number of materials increases from Lesson 2 through Lesson 9 and then remains the same for the remainder of the unit.

By adding materials strips to the poster when indicated, you will help students manage their science pail contents. The **Preparation** section of each lesson will tell you which materials strips to add to the poster that day.

Note: If you laminate the poster pieces, you will be able to reuse them.

Materials

1   poster board, 56 x 71 cm (22 x 28")
1   scissors
1   roll of tape

Procedure

1. Copy the blackline masters on pgs. 195–202.

2. Cut out and trim each materials strip. Cut out the title, arrow, and science pail.

3. Place and fasten the title, arrow, and pail on the poster as shown in Figure C-1.

4. You will use this poster for the first time in Lesson 2. Explain the function of the poster to the students at that time.

Figure C-1

PAIL

Empty.

1 PIECE OF BLACK PAPER

2 PAIRS OF GOGGLES

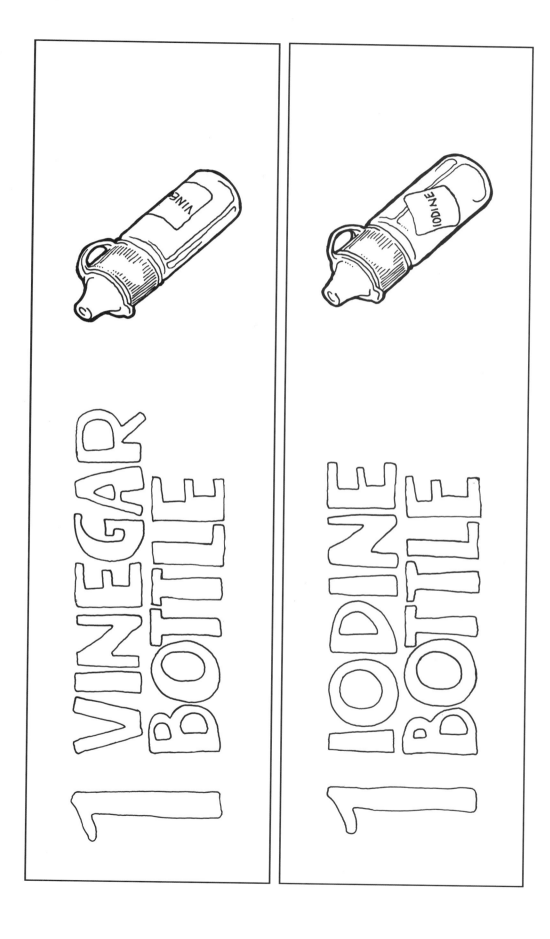

1 VINEGAR BOTTLE

1 IODINE BOTTLE

5 JARS OF UNKNOWNS

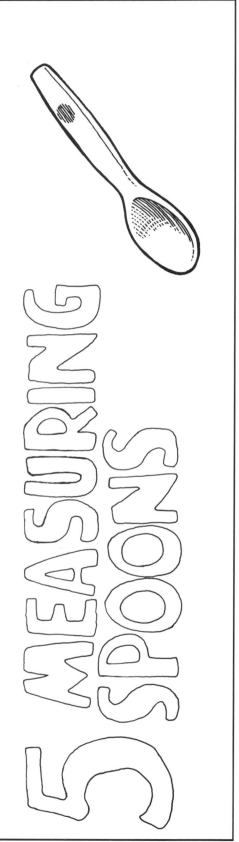

5 MEASURING SPOONS

2 MAGNIFYING LENSES

1 WATER BOTTLE

Making the Test Solutions

Red Cabbage Juice

Cut a red cabbage into eight parts and put them in a nonaluminum pot. Add enough spring water to cover the cabbage and boil for about 10 minutes. Pour the contents of the pot through a strainer into a container. Discard the cabbage leaves.

Cool the red cabbage juice and store it in a covered container in the refrigerator until needed. If you plan to store the juice for more than a few days, freeze it.

Sometimes the color of the cabbage juice will change slightly during storage. This will not interfere with the juice's indicator properties. If your cabbage juice becomes pink, however, you may wish to add a little base (e.g., ammonia). Similarly, if the juice becomes green, you may wish to add a little acid (e.g., vinegar). Or you can make a new batch of red cabbage juice.

Iodine Solution

Iodine may be purchased from a pharmacy. The "tincture of iodine" sold in the pharmacy is typically a 4.4% solution, which should be diluted for use in the classroom. (The iodine labeled "decolorized" will not work as an indicator.)

To make the diluted solution needed in Lesson 8, mix 10 ml (⅓ oz) of tincture of iodine with 240 ml (approximately 8 oz) of water. Since light will cause the solution's gradual degradation, store it in a dark bottle.

Ammonia Solution

To make the dilute 2% solution for Lesson 15, mix 5 ml (⅙ oz) of unscented, clear household ammonia with 240 ml (approximately 8 oz) of water.

Alcohol Solution

To make the dilute 2% solution for Lesson 15, mix 5 ml (⅙ oz) of rubbing alcohol with 240 ml (approximately 8 oz) of water.

Detergent Solution

To make the detergent solution for Lesson 15, mix 1 tsp of Super Washing Soda™ Detergent Booster with 250 ml (about 8½ oz) of water.

Observing and Describing Activities

The following activities will provide students with further practice in observing and describing objects.

1. Ask each student to bring an apple to class and to observe and record as many properties of that apple as possible. Then proceed as follows:

   ■ Place all the apples in one pile. Ask each student to remove his or her apple from the pile and then to put it back.

   ■ Have one student describe his or her apple to the group. Ask another student to find that particular apple in the pile on the basis of the description alone.

   ■ You may wish to repeat this activity several times. Following are some questions for discussion:

      ▪ How were you able to locate your apple? How were you able to locate a classmate's apple?

      ▪ How were your observations different from those of other students?

      ▪ What kinds of observations are most useful for identifying objects?

2. Go on a class scavenger hunt.

   ■ Choose a hunt site, such as your classroom or the playground. Divide the class into teams of two.

   ■ Have each team choose an object in the site and write a description of it.

   ■ Ask teams to exchange descriptions. Then ask them to locate the object on the basis of that description alone.

   ■ Have teams share their findings and discuss what aspects of the descriptions (such as color, size, and shape) were most useful.

3. Bring in some black-eyed peas and distribute a cupful to each student.

   ■ Have each student put his or her peas on a tray and observe all of the peas at once. Ask students to look for ways in which the peas are alike and ways in which they are different.

   ■ Have students choose one difference and sort the peas by that property. Then have students repeat this activity for a similar property, for two properties at once, and so on.

   ■ Ask the class to choose one property by which to sort and graph all the peas in the class.

4. Take the class to an outdoor site that has a variety of plants and structures. Have each student choose an object and make a "Describing Book" about it. Ask students to observe the object closely and, without drawing, to do the following:

   - Write a poem (such as an acrostic or concrete poem) that describes the object.

   - Write a paragraph describing where the object was found.

   - Write a paragraph describing a day in the object's "life."

   - Write a haiku about how the object makes them feel.

   - Exchange books and try to guess what object is being described.

5. Bring in milk, food coloring, and dishwashing liquid and follow these steps:

   - Give every two students a cup of milk. Have them observe it and record their observations.

   - Ask students to add a few drops of food coloring and record their observations.

   - Have students add a drop of dishwashing liquid to the milk and food coloring.

   - Ask them to observe and describe what happens. How did the properties of the milk and food coloring change?

   What actually happened? The fat in the milk does not mix with the watery food coloring. But when the dishwashing liquid touches the milk, it breaks up the fat. The fat spreads out, enabling the food coloring and milk to mix.

6. Choose an environment to explore with your class.

   - Have each student choose one of these properties to explore: color, odor, texture, sound, motion, or pattern.

   - Ask students to record everything they observe in their chosen category.

   - Have students draw or paint a picture of the environment by illustrating that one property.

   Create a mural of the environment using all the students' illustrations. Then, have the class write a poem describing all the properties of the environment.